THE NAVIGATOR TRILOGY
BOOK ONE

ASTRONOMICAL MINDS

THE TRUE LONGITUDE STORY

THE ROYAL AFRICAN COMPANY & THE SLAVE TRADE
GREENWICH OBSERVATORY & THE LONGITUDE DISPUTE
EDMOND HALLEY & ISAAC NEWTON'S SECRET INVENTION
THE PARAMORE MUTINY & THE 1707 SHOVELL DISASTER
THE LONGITUDE PRIZE DECEPTION &
THE CHRISTOPHER WREN BLACKMAIL CIPHER

TE

First published 2007 by Samos Books
Broadford, Skye, Scotland IV49 9AQ
www.samosbooks.org

ISBN 0-9556439-0-3

Library Cataloguing in Publication Data

Gerrard, Edward

Astronomical Minds; The True Longitude Story

Key words:-
Cloudesley Shovell disaster; Isaac Newton; Edmond Halley;
Slavery; Sextant; Longitude;
Greenwich Observatory; Westminster Abbey ciphers;
Christopher Wren blackmail cipher.

A catalogue record of this book is available from the British Library

Cover design by Charlie Broughton

Printed by Lightning Source UK Ltd., Milton Keynes MK13 8PR

CONTENTS

LIST OF FIGURES

INTRODUCTION

This is a true story of discoveries and inventions which should have immediately made the world a safer place but failed to do so because of internecine disputes and dubious national security issues, which jointly resulted in the needless deaths of tens of thousands of men, women and children.

The unlikely thread that weaves its way through this book is the invisible track of our Moon. As it slips slowly backwards against the 'fixed' backdrop of stars, each fresh monthly orbit tracks a different path through the heavens to the previous and the cycle only begins to repeat itself after 223 orbits. Rather surprisingly this long sequence was first recognized nearly 3000 years ago and was originally used to frighten the life out of the populace by predicting eclipses, in order to plant crops at the right time and to sacrifice virgins usefully. The possibilities of using the cycle for navigational purposes or to explain the make-up of the universe was probably not given serious consideration at a time when everyone believed the Sun was extinguished when it splashed down into the western ocean every night. How it managed to re-light itself next morning and where to find a fresh supply of virgins were both of more immediate concern. However, comprehending the whys and wherefores of this peculiar 18+ year cycle suddenly became of immense scientific importance in the 17th century.

First, and of absolutely no interest whatsoever to 99.999% of the world's population in the 1660's, was the fact that the key to a better understanding of the universe lay in this decidedly odd orbital behaviour of Earth's only satellite. The elliptical paths of the six known planets were being forecast with ever-increasing success and a few exceptional brains were trying to explain why these orbits were not perfectly circular and why everything did not either spiral into the Sun or fly off into outer space.

However, determining longitude on the high seas was the second and urgent commercial reason for acquiring a full understanding of this lunar phenomenon. Fixing ones longitude far from home depended on being able to carry 'home port' time with you on your travels and the Moon's fleeting proximity to nearby stars could theoretically be used as the hands of a heavenly clock; one which would always keep perfect time (see Glossary for explanation). European nations were laying claim to distant territories half a world away that they could not defend because they were unsure of their exact location or in some cases, even how to find them again. Clearly anyone able to determine their position in the middle of nowhere would gain a powerful advantage over rivals.

The longitude holy grail had two other potential 'timer' candidates, both of which appeared easier to achieve than the hugely complex task of publishing advance prediction tables (*ephemerides)* of the Moon's exact position in the sky at any given time. A reliable mechanical clock was one obvious solution and using Jupiter's orbiting moons as an alternative heavenly clock was the other (chapter 2). A reliable timer of any kind would enable sea-borne commerce to blossom and, sadly of less importance, huge numbers of poor 'savages' might extend their miserable lives by a few years through being transported safely across longitudes over the mid-Atlantic (the dreaded 'middle-passage') from 'old' Africa to the 'new' Americas.

A brief reading of the Notes and Glossary sections at the back of this book may be helpful.

CHAPTER 1

Slavery, the disgusting business of trading in people, is, like prostitution, one of the oldest professions in the world. The Arabs and Moors of North Africa and the Middle East were already past masters at the art of capturing and selling African Negroes for profit long before European nations began participating in this potentially lucrative trade.

European involvement started more or less by mistake soon after Prince Henry of Portugal had established a navigation school for sailors on a wild and windswept site at Sagres on the Cape St. Vincent peninsular overlooking the Atlantic Ocean. His intention was to teach budding pirates how to safely venture south into the tropical waters of West Africa to trade or steal anything of value they might come upon; which they did very successfully. Hippo teeth, elephant tusks, exotic timber and above all else gold, soon became popular items on their shopping lists although because a certain loss of face resulted whenever a captain returned home with anything less than a handsome profit, a degree of double-dealing with the native suppliers became the norm. A few glass beads or a knife in return for a pouch of gold nuggets and if the native was not alert, the knife was in his back and not his pocket.

In 1442, following a deal of this type that backfired, the Portuguese captain of a caravel inexplicably acquired *'10 blacks, male and female'* instead of the elephant tusks he had hoped for. He reluctantly took his human cargo back to Lisbon and saved face by, with some difficulty auctioning the Negroes on the dockside.

Eight years later Pope Nicholas V brought the Vatican into this dirty business with a vengeance. He blithely authorised the Portuguese to reduce to perpetual slavery any non-believers captured in 'their' West African territories. With God's earthly representative approving the capture and sale of non-believing Negroes (who were, incidentally unlikely to be given the option of converting to Christianity) any niggling moral doubts were removed and Prince Henry's navigators now lacked only a ready market. But there was little chance of finding a good one in Europe or for that matter the Near East where Arabs and Moors had maintained a stranglehold on the slave trade for centuries.

In 1492, three decades after the death of Prince Henry, Christopher Columbus surprised himself, the rest of his expedition and millions of Europeans by his discovery of the New World. Some had even expected his venture to end in disaster when his three ships fell off the edge somewhere out in the mighty Atlantic. Once the enormity of his discovery had sunk in, most of Europe's leaders simply accepted that the New World was offering booty and territory beyond their wildest dreams and set about grabbing as much as possible. They also knew that, once the local inhabitants had been subdued these new territories would have to be defended against other European claimants.

A major problem occurred at the very outset because Columbus thought he had discovered somewhere else. Neither he nor his sponsors had any way of checking *how far west* he had actually sailed, although Columbus did know the approximate latitude of his discoveries, and the precise longitude of those discoveries was to remain uncertain for over 200 years. How can a nation lay claim and then defend a territory whose position is not defined?

In an effort to side-line this problem, the enormously powerful Catholic Church made full use of papal treaties to split the world in two from pole to pole, top to toe right down the centre of the Atlantic. The new Spanish pope, father of the infamous Lucrezia Borgia, gave trading rights to any undiscovered lands in the 'new' western half to Spain. The 'old' eastern half was given to a less than grateful Portugal, most of which the small country had already laid claim to anyway. Claims of other nations were airily dismissed out of hand and Pope Alexander VI then busied himself with more important matters such as parading round Rome with his latest mistress on sunny afternoons.

Unfortunately for Spain, the imaginary central Atlantic dividing line did not make due allowance for the eastern portion of undiscovered Brazil which stuck out into the south Atlantic so far that even the most primitive methods of determining longitude confirmed that it was in the African Portuguese half! Which is how Portugal, much to the pope's annoyance, gained a foothold in the Americas alongside its powerful neighbour Spain.

With the discovery of Brazil, Portugal suddenly found itself uniquely placed; it held the key to the American labour shortage. Not only did the Portuguese hold the trading rights to West Africa with its apparently limitless supply of Negro slaves, but there was land on the other side of the Atlantic that they now also 'owned' where they could make very profitable use of all this cheap labour; and they did. Not only that, but Portugal could also sell any surplus slaves to the labour-hungry Spanish American colonies; they did that too. The Spanish pope's double-dealing had failed miserably and the trans-Atlantic nightmare had begun.

As early as 1501 petrified West African slaves, many of whom had never tasted salt water, were being shoe-horned into the holds of ships and transported across the mid-Atlantic in their thousands by the Portuguese both to their own Brazilian lands and to the Spanish colony on Hispaniola, the island which is now divided in two and known as Haiti and the Dominican Republic. The sugar-cane plantations of Brazil and Hispaniola were about to be cheaply and rapidly developed using Negro slave labour almost exclusively. The American natives continued to decline by the hundreds of thousands, partly because of heavy losses incurred through fighting the well-armed and mounted invaders, partly because of suicides, partly through introduced European diseases but ironically mainly because of imported African diseases such as yellow fever, smallpox, virulent strains of malaria and hookworm. The native population of Hispaniola was estimated to have been at least 200,000 when the island was discovered by Columbus. This was reduced to less than 1,000 within 50 years.

In 1514 an enlightened Pope Leo X denounced slavery and the rapidly expanding slave trade was officially brought to a skidding standstill; but not for long. Within two years Spain quietly began granting licences to third parties to sell slaves to its American colonies; not because Portugal was abiding by the pope's worthy desires but because the Portuguese were continually giving priority to their own mushrooming colonial requirements. God's wishes would have to take a back seat so long as overseas Spanish plantations were crying out for labour.

Within a decade the Negro self-styled king of the Congo was desperately and ineffectively petitioning King John III of Portugal, protesting that despite his being a devout Christian, Portuguese slavers were depopulating his country. So much for Christian immunity and the Vatican's embargo; but his plea received little sympathy because the Congolese king had been happily selling his subjects into slavery for years and was only now complaining because Portuguese mercenaries had taken matters into their own hands. These enterprising blackguards had started slashing their way into the bush, catching their own slaves and brazenly refusing to pay any commission to the king!

Sometime early in 1555, a native South American captive revealed the location of a small silver mine to his Spanish inquisitors. The Potosi (Bolivia) mine would soon be yielding truly immense quantities of silver, hacked from a mountain in terrible working conditions by Negro slaves. The semi-precious metal became devalued, the demand for slave labour increased, more silver found its way to Europe, currencies were undermined and for a time inflation became the norm. The Spanish also discovered that shipping this treasure back to Europe was not easy. The slow-moving, poorly navigated but heavily armed Plate fleets attracted a great deal of unwanted piratical attention from the moment they set sail until they docked in Spain. Even then they were not always safe.

Fifteen years later demand for Negro slaves finally outstripped the supply. This had been caused by the amazing popularity of sugar in Europe, by yet another smallpox epidemic in Brazil and because the Negro population of West Africa had been decimated. Slave suppliers were now forced to travel deeper into the dark recesses of central Africa to capture their human quarry.

At the dawn of the 17th century hundreds of thousands of Negro slaves were being worked, literally, to death on plantations in central and southern America. How many had died on the ever lengthening overland journeys before even reaching the fortified holding pens on the coasts of West Africa, during the terrible trans-Atlantic crossing, or through disease after arrival is anybody's guess. No one kept proper records, and so long as the losses could be replenished, no one much cared.

To date, African slaves had been used only in central and southern America but in 1619 a Dutch ship hijacked the cargo of a slaver on the high seas and bartered this shackled line of humanity for food and water at Jamestown, Virginia. The slave trade had spread to God-fearing English colonial North America, although in fairness to the people of Jamestown this first batch of Negroes was treated with relative kindness. Oh that further shipments to North America would be treated so humanely.

In 1640, Portugal allegedly upset its overpowering next door neighbour and Spain promptly cut all commercial ties, forbidding its merchants to purchase slaves from its long-time supplier. Between 1640 and 1662 there were no official slave transactions between the two countries, but the demand for labour in the Spanish American colonies was greater than ever. Just as Spain had intended, there were plenty of others willing to fill the void and the Portuguese in 'their' West Africa suddenly found themselves in the middle of a vicious tug-of-war over supplies of Negroes, a war that they stood little chance of winning. The sheer weight of the enthusiastic opposition would see to that.

First came the Dutch flying the colours of 'The Dutch West Africa Trading Company' followed rapidly by several small French syndicates and the English 'Guinea Company'. All intent on making inroads into Portugal's monopoly in human misery now she was no longer under the protection of the powerful Spanish; foreign competitors who were determined to buy slaves for the least possible amount in order to fill Spain's order books in return for payment in silver from the Potosi mine. Any slave surplus could go to meet the booming demand of their own country's American colonies; a demand now being boosted by the latest European fashionable demand for tobacco and chocolate.

This in turn again increased the need for capturing and transporting African Negroes. A tragic merry-go-round, fuelled by greed that was spinning rapidly with only the Devil himself in charge. But the Devil, with

Spain's assistance had just played a trump card which now set the merry-go-round spinning faster still.

Originally Portugal had shipped slaves to the Americas in Portuguese ships manned by expert seamen wise in the ways of the dangerous Atlantic Ocean. Wise through a hundred years of collective experience handed down from one generation to the next and guarded as securely as any crown jewels. But the new players, the Dutch, French and English, had no such data at their disposal. Greedy captains were setting sail in the hurricane season or becoming becalmed in the doldrums without adequate water supplies, running into rocks they were unaware of, or simply losing their way. And the ultimate losers once more were the poor slaves shackled together and stacked onto shelves below decks. Slavers that ran out of water simply threw the Negroes overboard, all too often still chained together and whilst captains who sank their ships sometimes escaped with their crews in open boats; not so the manacled slaves.

Departure from the Iberian Peninsular south to West Africa, then across to the Caribbean or Brazil and back across to Iberia had involved little but constant down-wind sailing, taking full advantage of favourable trade-winds and surface currents. A triangular passage that, outside the hurricane season rarely encountered stormy conditions. But when, by necessity latter parts of the third leg of the triangle forced ships much further north on their passage back to Dutch, French, English or Scandinavian home ports, heavy-weather sailing became the norm rather than the exception. Navigating in bad weather in contrary sea conditions required expertise none possessed. With inadequate charts and no way of gauging longitude on this more northerly third leg, shipping losses mounted alarmingly. Although ships were cheap to manufacture and slaves and seamen dispensable if needs must, this third and final leg of the triangular passage was bringing home the real profits, the valuable commodities produced so cheaply by slave labour in New World colonies.

By the middle of the 17th century even the most ignorant of Europe's rulers had realised that the mushrooming fight over the New World's riches depended for success on navigational expertise, which despite all efforts had not advanced much since the days of Columbus. Any successful battle for command of the open ocean and defence of new overseas territories required navigational prowess of a much higher level than that needed for sailing within the enclosed Mediterranean Sea, or even up and down the west coast of North Africa. The merchant fleets were in even more urgent need of a navigational upgrade than the lumbering cannon-toting naval battle wagons if they were to fully exploit the riches of the New World. The vast Atlantic Ocean was not simply a bigger version of the Mediterranean;

it was a deadly dangerous place as they were discovering to their cost.

Which is of course why all the European seafaring nations were backing a new breed of citizen, the scientist, in their efforts to improve the age-old methods of navigation at sea. Being greedy they all went for the jackpot; they became embroiled in the 'quest for longitude at sea'. If they did but know it, akin to trying to land on the Moon at a time when the Wright brothers were still manufacturing bicycles.

The desperate search for finance by the re-instated House of Stuart had commenced the moment Charles II replaced his late, unfortunate father as King of England in London in 1660. One of Charles' first actions was to plead successfully with his parliament to tighten further the old English Navigation Act of 1651 which had banned the import of goods into England or *English colonies* unless carried in English-owned vessels and had, incidentally started a war with the Dutch.

Using bribery, coercion and the vociferous support of London shipping merchants, Charles had been so successful that by 1663 English maritime law was boldly proclaiming that most of England's colonial produce could only be sold *to* England and that all *foreign* goods shipped to England's colonies had to do so via English ports. Charles had somehow managed to instil into his members of parliament the vision of England clambering rapidly to very pinnacle of world power on the back of its merchant navy. He personally merely hoped his private treasury would at last begin to fill.

This was not a happy state of affairs for the colonial producers making vast profits through the exploitation of slave labour because this meant that their new king now had control of, and could raise taxes on, these fortunes being made in his American colonies. Which of course he immediately set about collecting and spending as fast as possible.

James Duke of York, younger brother of Charles II and Lord High Admiral of England's powerful but inefficient and under-funded Royal Navy, now made a second play on behalf of the monarchy with a brilliant masterstroke. James took a major stockholding in a new trading company and persuaded his mother, sister-in-law and siblings to subscribe. They in turn advised many of their upper crust friends to do likewise. Significantly there seems to be no record of his skinflint brother Charles paying his promised £6000 (approximately £1,200,000 at today's values) into the kitty.

In 1663, the Duke of York's new and misleadingly titled 'Company of Royal Adventurers Trading into Africa' (no royal stockholder in the company ever set foot on African soil, let alone ventured inland) managed to obtain a unique charter from parliament in highly questionable circumstances which gave the company what appeared to be a monopoly on the entire West African trade. The 'Guinea Company' of merchants and other English traders in West African produce were put out of business overnight and CRATA, a private company, became the only organisation in the world legally permitted to sell West African slaves to English colonies in the Americas. Because their captains would be swapping the slaves for colonial produce which they could then, by law, only deliver back to England, vast profits for the stockholders appeared assured. In theory the king's coffers would benefit twice; CRATA dividends and import taxes.

At this point in the story gold becomes almost as big an issue as Potosi silver. From Henry the Navigator's time, gold, ivory and hardwood had all been exported from West Africa and now the CRATA prospectus highlighted these products in order to attract more investors. Indeed gold nuggets, exchanged for trinkets, were one of the first items shipped back to England by the company. Some of the rarest English coins, the 1664 Golden Guineas of Charles II were minted from this 'Guinea' gold and were stamped with the CRATA logo, an elephant. Although slave trading would play a leading role in the company's fluctuating fortunes over the years, underhand double-dealing in gold would eventually bring the company and its successor to their knees.

The seed money the company managed to attract was insufficient to build their own fleet of trans-oceanic vessels because many subscribers took a leaf out of their sovereign's book and also somehow forgot to pay in full for their stock. So the company had to charter armed merchantmen from other English ship-owners, now conveniently desperate for business.

All started well enough and in seven short months following the end of the 1663 hurricane season, vessels chartered by CRATA actually delivered 3,075 Negroes to one single island, Barbados. There is no record of how many were originally purchased in West Africa, only of how many arrived alive and in saleable condition. In the same year the company also contracted under licence to sell 3,500 slaves a year to colonial Spanish interests but very few were officially delivered because of dubious book-keeping methods and ongoing international competition in West Africa.

But the poor wretches were still being fought over by powerful foreign trading companies who refused to recognise the monopolistic rights of CRATA and forced the company to compete for the slaves it needed to fulfil its contracts. The fights were not based commercially on the highest hard cash bidder getting the item, but by who could offer the brightest brass pots or most effective firearms. Indeed many English captains were forbidden to offer cash for slaves and a burgeoning industry manufacturing trade goods back in England owed much of its future success to the Duke of York's fledgling business venture.

Sadly the fight for live produce often continued on out to sea. Slaver captains took a leaf out of the Skua's book, a predator seabird they were all familiar with, and, just like the bird, they evolved clever tactics to steal someone else's cargo. So the design of slave ships was soon transformed into armed fast frigates carrying as many as 30 cannons; ships that mimicked the designs of the English Royal Navy's workhorses, but had removable interior deck levels for packing in layers of slaves.

The proud Dutch in particular resented their trade being restricted by the latest English Navigation Acts. They also took a dim view of their ships being attacked on the high seas by heavily armed frigates chartered to a private English trading company. They took an even dimmer view of being plundered in peacetime by units of the English Royal Navy which were, so they believed, acting on the direct orders of the Duke of York who was, as the Dutch knew only too well, also a major stockholder in CRATA.

However, and unfortunately for the investors, the company's presence in West Africa was simply not well enough organised to take advantage of their unique position. Neither was the Royal Navy who appeared to the CRATA rank and file stockholders to be little better than a bunch of pirates unable or unwilling to properly protect England's merchant fleet. When an English slaver did manage to legitimately negotiate for, and successfully load a consignment of human misery, all too often the entire caboodle was promptly hijacked by a heavily armed Dutch privateer, whose captain held the crew to ransom, changed the flag, put a skeleton crew on board and sailed the ship off to any non-English New World port which was willing to purchase the cargo.

The paramount reason for the English Royal Navy becoming embroiled in these unprovoked attacks on foreign merchant shipping was purely financial. The seamen were, if at all, poorly paid (chapter 4). Everyone from Admiral of the Fleet on down to the lowliest deckhand or powder monkey relied on the proceeds of prize money and a precise sliding scale was strictly enforced. However the chances of getting one's head or legs knocked off in a major naval engagement far outweighed the likelihood of prize money.

Not so the armed raider attacks on merchantmen and one successful attack on a slaver could net an ordinary seaman the equivalent of a year's wages, wages he sometimes had to wait years to receive. Mimicking a Skua was a lucrative business just so long as the marauders did not stray too far out to sea and become lost in the inhospitable reaches of the Atlantic.

CHAPTER 2

Most mariners could use a magnetic compass to hold a steady course and could determine an approximate *latitude* by judging the height of the pole star but that was about all. Careful compilation of data on the Sun's seasonal changes (caused by the Earth's elliptical orbit and permanent tilt) permitted reasonably accurate solar advance prediction tables (*ephemerides* or *almanacs*) to be published in the latter half of the 17th century.

Measuring the continually changing noonday Sun height complemented the difficult pole star method of determining latitude, in part because the pole star was not actually positioned directly above the North Pole and partly because the horizon was difficult to identify at night. Although the latitude problem was, on dry land by now completely mastered, ships' navigators still experienced difficulties in measuring solar angles from the rolling deck of a vessel on the high seas. Using the standard angle-measuring tool, the back-staff (figure 1), required the observer to view the horizon to the fore whilst trying to manoeuvre the Sun's shadow onto a slit in a small piece of wood from over his shoulder, a degree of skill few could master.

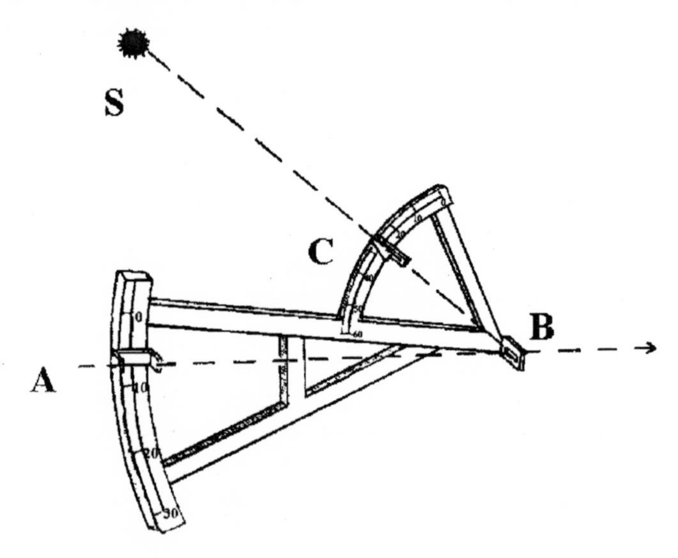

Figure 1. The Back-staff.

The observer stood with his back to the Sun looking through pinhole sights (A) and slit (B) at the horizon. Sight (A) was adjusted until the Sun's shadow touched the slit (B), providing its altitude.

More efficient than the earlier astrolabe and cross-staff used by Columbus, but at very best and in calm weather, latitude could only be gauged with certainty to within about 15 miles or ¼° (about ½ the apparent diameter of the Sun). Even so, sensible navigators were unwilling to put too much trust in back-staff measurements because they had no way of knowing if the instrument had warped in the heat or humidity, or had been roughly treated during the voyage; no way of checking against a known standard.

One obvious solution to the far more difficult *longitude* problem had attracted the regular attention of philosophers and inventors since the time of Pythagoras of Samos. Because the Earth completes one spin every 24 hours the Sun appears to travel once round the Earth in that time; the Sun moves 15 degees across the sky (360° divided by 24) every hour. If it was just reaching its zenith at Alexandria in Egypt it would not do so 15° west until exactly one hour later. If you could somehow carry Alexandria (home) time with you whilst you sailed west all the way to the island of Sicily you would discover the Sun reached its zenith about an hour later than it had done at home; the locator longitude would be about 15° west of Alexandria. Not easy to achieve in ancient times because your time candle or water clock was not very reliable. Which is one reason why the length of the Mediterranean was still not properly established at the beginning of the 17th century. It was a very good idea, but impractical because no one could make an instrument capable of retaining home port time during a long sea crossing.

Retaining the essential 'home time' using candle or water clocks whilst crossing the Atlantic to the New World was clearly out of the question. This explains why Christopher Columbus had not the foggiest notion of how far west he had sailed (his longitude) on that first pioneering venture, yet knew to within a degree or two how far north of the equator he was (his latitude). His arrival point in the New World was somewhere in a box of about 80 miles in 'height' and well over one 1000 miles in 'length'. Nevertheless Columbus already knew something of the eclipse part of a heavenly clock method of assessing longitude; lunar eclipses occurring more or less at the same time whatever the location. But knowing of the theory was not the same as making practical use of it and a later attempt to use an eclipse to establish the longitude of his discovery was in error by a massive 23° and did nothing to shorten the length of his New World box.

Two hundred years after the Columbus discoveries, voyages across the Atlantic were still being monitored by utilising the primitive boat-speed method; by keeping a careful tally of daily 'distance run' figures. This was computed by occasionally throwing overboard a knotted line with a wooden float on its end and counting how many knots had been paid out before the sand in the top half of the (egg) timer ran out. This information was then compared with data collected from previous voyages and an approximate arrival date worked out. Once the expected journey time was nearing its end, extra lookouts were posted, the depth of water was constantly checked with a lead-weighted line and at night most sails were furled. All in all, not a very secure method of avoiding running into a hard lump of land or becoming trapped on a lee shore in the dark.

But vicious international competition and continuing en-route losses were now threatening all the trans-Atlantic slave traders and their backers with bankruptcy. A scapegoat was urgently needed and the blame, rightly or wrongly was placed squarely on the continuing failure to solve the 'longitude' problem (Glossary).

By the late mid 17th century, very fine clocks and watches were being produced; surely someone could make one that could be used at sea? A reliable time-piece of some kind that allowed the navigator to carry that original home (London) time with him on his travels around the Atlantic would let him know the Sun was indeed reaching its zenith two hours later than his clock indicated and he must be 30 degrees west of London. But what if the clock was unreliable? With every four minutes of clock error comes 1° of longitude error, and how can the navigator know that his clock is keeping good time? Better than using a water clock or sand timer maybe, but ....

What of the lunar clock? This method of determining longitude was clearly never going to be easy to achieve. A sailor would have to know all the details of the Moon's position against the backdrop of stars *in advance;* lunar almanacs would have to be published. Unless the target star was very close to the Moon, (or an eclipse of some sort was in progress) any angular measurements would need to be very accurate. A tiny two minutes of a degree (or one fifteenth of the Moon's diameter) error in measuring the distance between a rim of the Moon and a star from a wildly rocking boat could result in a longitudinal error of a full degree; 60 nautical miles at the equator, even if the prediction tables were perfect. It would also be imperative that the navigator was not exhausted and could wedge himself in somewhere dry in order to complete a couple of hours of complex error-free mathematics.

Although astronomers had been aware of the Moon's erratic movements for thousands of years, in 1660 they still did not understand *why* it behaved so oddly. This conundrum had to take its place in the lengthening queue of astronomical puzzles. Certainly no one knew anything of gravitational effects on the Moon's orbit exerted by other heavenly bodies or if Earth's own elliptical orbit was regular; even its polar circumference had only recently been properly confirmed. Currently the English polymath Robert Hooke was involved in a priority dispute over who first suggested the Earth was turnip shaped. If it was fatter round its middle (and it is), this too would upset those oh so delicate calculations.

Without a full understanding of all these points and several other factors, satisfactory lunar almanacs could never be published. It is easy to forget that only 50 years previously Galileo Galilei had escaped death by a cat's whisker for daring to suggest that the Earth was anywhere but slap bang and unmoving at the very centre of the universe, with every other lesser item, including all the stars in the heavens, acknowledging her might by dutifully, if somewhat speedily, orbiting her in a series of perfect circles once a day. He had found himself in this papal trap because he had discovered another method of determining longitude.

The (re)invention of the telescope in 1608 by the Dutchman Hans Lippershey had enabled Galileo, a little more than a year later to discover four tiny moons orbiting the planet Jupiter. These buzzed round the planet at predictable speeds, sometimes disappearing behind Jupiter and then reappearing again on the other side as if by magic. It became painfully obvious to anyone possessing a telescope that at least four of God's creations were misbehaving. They were all orbiting something else!

Johannes Kepler, Imperial Mathematician of the Holy Roman Empire, quick off the mark on hearing of Galileo's discovery and understanding its navigational importance, wrote an open letter of support entitled *'Conversations with a sidereal messenger'.* Kepler and Galileo had both realised that the 'stars' of Jupiter could be used for the same purpose the Moon and true stars could be used; as the hands of a heavenly clock in order to determine longitude (the sidereal messenger). When one of Jupiter's moons suddenly disappeared from view behind its mother planet that event was observed at almost exactly the same time wherever the observer was located on the Earth's surface. This heavenly clock would surely prove far easier to predict than the motions of the Moon and both astronomers were absolutely right.

Galileo attempted to cash in on his discovery by claiming a large pension from the Spanish in return for furnishing them with the details. He tried more than once but the Spanish were not impressed; at least if they were, they never admitted it. In reality the Spanish and everyone else did eventually make use of the Jupiter's moons method, but only as a means of assessing longitude on dry land, where proper timed observations of the tiny moons could be viewed through a very long telescope from a rock-steady platform. In this manner did the Spanish, much to their delight, many years later discover that they had sold to the Portuguese the trading rights to the Spice Islands way round the other side of the world, rights which under the papal-induced Treaty of Tordesillas that had split the world into two halves, the Portuguese had actually 'owned' anyway.

The Dutch who were also desperately seeking a method of determining longitude on the high seas were a little more respectful and awarded Galileo a gold chain in lieu of their full prize of 30,000 florins. In 1638 when this award was to be formally presented, by now ill and blind, he declined to accept it, being still under house arrest because of his heretical publications. The last defiant gesture of the man who had discovered the solution to the basic longitude puzzle. However Galileo had always imagined that it would be a relatively simple task to produce almanacs which would enable a navigator to know in advance to within a few seconds exactly when a Jovian event would occur. He was wrong and for the next 60 years the only way to use this method to determine longitude on land was for the astronomer in the field to compare notes with the base observatory when he returned from his expedition.

Primitive Jovian almanacs were already in existence in 1660; they were inaccurate for several reasons as will be explained in chapter 9. But as the Spanish and Dutch had discovered, for the seaman navigator there were major snags even if the almanacs had been reliable, as anyone who has tried to focus on Jupiter (let alone one of its moons) with a pair of binoculars from the deck of a yacht will confirm. So although the puzzle might, in theory now have been solved for all dry land sites once the timing of the moons could be nailed down properly, sailors were only a little better off whilst on the high seas. At least now countries and boundaries could be accurately mapped and positioned and the sailor would surely soon know, for example where the coastline of Brazil was positioned on the world globe. Or at least one would think so, but to plot such important locations meant landing first. The Portuguese, because of their accumulated knowledge of voyage times and prevailing winds already knew how to get to Brazil without being able to determine longitude. They were therefore suspicious of foreigners coming ashore carrying 25 foot long telescopes, cumbersome tripods and large pendulum clocks. So touchy were they that foreign vessels arriving in Brazilian ports for any reason other than approved trading were likely to be blown out of the water, or at best impounded.

The Jovian land-based method of assessing longitude 'only' required a firmly based telescope, an almanac of Jupiter's moons, a fixed and leveled quadrant to measure the height of the Sun or pole star to obtain local time for comparison, plus a pendulum clock. But either of the two age-old contestants for determining longitude at sea (mechanical clocks or lunar heavenly ones) also required an instrument capable of measuring angles very accurately from the rolling deck of a ship. That instrument most certainly was not a back-staff.

CHAPTER 3

Robert Hooke, surveyor, architect, astronomer and mechanical genius was born on the Isle of Wight in 1635. His father was a church minister and almost certainly committed the unpardonable sin of taking his own life in 1648 after being persecuted and heavily fined for helping Charles I whilst the monarch was sheltering from his enemies in nearby Carisbrooke Castle. Young Hooke's tutor on the island, William Hopkins gave the king lodgings at one point and also later paid a heavy price for doing so. There is no evidence to suggest that Charles' eldest son ever met Hooke junior and certainly none that he attempted to repay any debt, real or imagined. The nearest Hooke ever came to being presented to Charles II was on a couple of occasions when the monarch broke his promise and failed to turn up at functions Hooke had been specially preparing for.

Hooke, by all accounts was a sickly baby not expected to live, and seven years later his pessimistic mother apparently still rated his chances of survival only a little above zero. Book study gave him headaches but the curious inner workings of clocks in his house and the near miraculous power of pulley blocks in use on sailing ships in the local harbour helped nurture his innate flair for mechanics. The stratified sea cliffs and fossil beds near Freshwater set the over-active mind in its frail frame thinking furiously and he came to the remarkable and heretical conclusion very early in life that God was unlikely to have created the world in a week six thousand years ago. This mentally hyper-active child also discovered he could master any number of skills simply by observing craftsmen at work, which was fortunate given his aversion to traditional teaching methods.

Immediately following his father's tragic death the 13-year-old took his small inheritance and set off to London in search of fame and fortune, first as an apprentice to a portrait painter and then, because the smell of paint also brought on headaches, he used some of his cash to enrol at Westminster School. There Hooke developed interests in mathematics, learned Latin and Greek; his headaches clearly only manifesting themselves selectively. Unfortunately he also developed curvature of the spine, a handicap even his phenomenal mental powers could not correct. At about this time he began informing anyone willing to listen that he had invented many different ways of flying; a latter day Leonardo da Vinci in the making, although when pressed he declined to go into details claiming someone might steal his ideas. This tendency to brag outrageously on occasions would all too soon become highly damaging to his and others' scientific careers.

Hooke was at Christ Church Oxford between 1653 and 1658 where he became friendly with fellow pupil Christopher Wren; a friendship which was to last, unusually for him, most of his life. His years at Oxford University saw his raw youthful natural talent mature into brilliance and he was eventually rewarded with a Master of Arts degree. Two recent Hooke biographies provide an insight to his abilities (and shortcomings) in their titles; *'London's Leonardo'* and *'The Man Who Knew Too Much'*.

Since its inception in 1660 and the granting of a royal charter in 1663, the Royal Society, of which Wren was a founder member, had granted fellowships to the wealthy and influential as readily as to those with scientific talent. Regular demonstrations were usually organised by Hooke, the society's Curator of Experiments; exhibitions which would nowadays sometimes be considered little more than shows pandering to the ghoulish morbid curiosity of unemployed gentry who had nothing better to do on wet afternoons. Details were duly noted in the society's journal book, sometimes in minute detail but more often not; unfortunately mingling nonsensical reports with scientific nuggets of gold the purity of which was difficult to assay when essential data were lacking. Correspondence from overseas fellows announcing new discoveries and usually written in Latin were also often lacking detail.

The RS meeting room in London's Gresham College was not the only place where exciting new scientific developments could be discussed. Coffee houses were noisy popular London meeting places where not everyone merely drunk coffee, smoked tobacco and passed the time of day reading news sheets and broadsides. Because Hooke was a lonely bachelor with a lively mind and the need for stimulating debate, he spent every spare evening in one or other of at least 60 of the hundreds of coffee houses within and without the thick boundary walls of the noisome city of London.

He even had his own reserved corner table at Garraway's in Exchange Alley where anyone willing to listen or to engage in lively logical debate was made welcome and often Hooke would organise scientific demonstrations for customers rather in the style of magic shows. In some ways establishments like this took the place of the universities where in the 17th century students rarely bothered to attend lectures, let alone discuss anything verging on the academic. Isaac Newton was to deliver the first of his revolutionary lectures on the world of physics to a very sparse audience at Trinity College Cambridge and his second lecture was delivered to an entirely empty hall. He could have filled a room at Garraway's any day of the week, just as long as he was prepared to put up with Hooke's sarcasm and debate half formulated ideas which, when properly nurtured in private, were to lay the foundations which would one day enable the United States of America to land men on the Moon.

But if the sparse written notes of the RS were sometimes unhelpful, there were major flaws with coffee house scientific verbal pronouncements. Apart from Hooke's spasmodic personal diaries and a few surviving letters, no record of any of his coffee house discussions or demonstrations were made, which later led to acrimonious priority disputes. To add to the confusion those unwilling to use either the RS or a coffee house for proclaiming their discoveries often resorted to enciphered notes in order to register their initial claims. This clever ruse conveniently gave nothing away, but allowed details to be fine-tuned prior to official publication (chapters 6 & 24).

One of the most interesting topics discussed regularly at Garraway's was naturally enough, the long-running quest for discovering a method of assessing longitude at sea. Why was a sea-going timepiece so difficult to construct when good land-based clocks were already being produced? The problems Hooke ran into during his efforts to solve this part of the puzzle over an extended period between 1660 and the beginning of 1664 serve to illustrate some of the snags he came up against.

First he invented an entirely new watch escapement mechanism which used a novel leaf spring design. This improved accuracy, but became unreliable when moved out of an upright position, as did the existing very accurate pendulum-driven clocks designed by the Dutch horologist and RS fellow, Christiaan Huygens, which were based on Galileo's original principle.

To counter this basic flaw, after much research and wig scratching Hooke then designed a pair of spring-driven balance wheels which would absorb the rocking and shocks of the ocean; in truth by far and away the nearest anyone had yet come to producing a reliable sea-going timepiece.

Probably sometime in 1664 Hooke showed this watch to the president of the RS Lord William Brouncker and influential fellow Sir Robert Moray, initially concealing the inner secrets from them. Matters got as far as drafting a proposal for the king to consider, a proposal which contained an offer to divulge the secrets of the mechanism in return for the granting of a royal patent. But because of a clause inserted by the king's advisers which would allow others to benefit should they improve on Hooke's design, negotiations collapsed and Hooke stamped off in a huff before his invention could be properly tested; or even examined.

Somewhat incautiously, Hooke boastfully revealed some of the secrets at a lecture he gave at Gresham College (chapter 6). Following the collapse of the patent application the entire project came to a screaming halt and

Hooke's grasshopper mind jumped across to a second part of the longitude solution he had been working on spasmodically.

Hooke's surveying talents had made him something of an expert on angle-measuring devices and in micrometer fine tuning to gauge such angles precisely; in theory if not always in practice. So the years of working on what were in effect navigational tools, together with the countless hours spent discussing the longitude question with Wren and anyone else willing to contribute, had given him a deal of genuine expertise in this field.

He designed a small portable contraption that used a mirror and a telescope to measure angles (figure 2). There is no evidence to suggest that this instrument was ever tested at sea and it was to remain gathering dust in the society's repository for the next three decades (chapter 13).

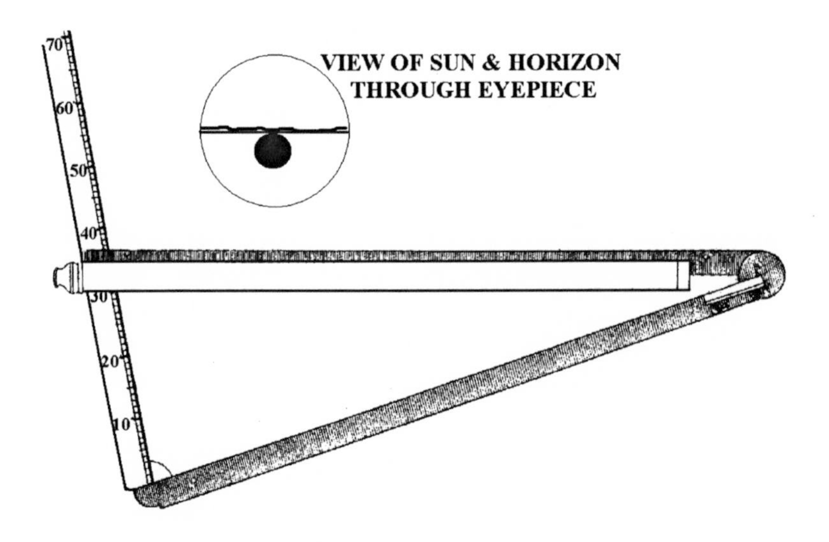

Figure 2. Robert Hooke's mirrored angle-measuring device.
The horizon is viewed through the telescope and lined up across the top of the diagonal polished solid metal mirror. The heavenly object's image is reflected down on to the mirror and bounced in through the telescope. The horizon and the (limb of) the Sun must both touch the edge of the mirror as well as each other; at which stage the angle was measured off along the sliding scale. The device was a theoretical improvement on the back-staff (figure 1). The view through the eyepiece would actually be inverted.

Hooke then constructed yet another ingenious device which apparently utilised twin telescopes and eyepieces and polished metal mirrors (specula) as a means of viewing Jupiter's moons from the deck of a ship. Hooke and Wren tested the instrument whilst bumping over potholes in a carriage but it didn't work too well, was dismantled and the telescopes put to other uses. This was now the second time the mirror idea had been employed in a prototype angle-measuring device which Hooke had invented.

Having dabbled in timers, angle-measuring devices and a method of utilising Jupiter's moons, all without much success, the only sector of the 'longitude quest' left undisturbed by Hooke was the lunar orbit question. But Hooke and Wren both understood that the Earth's own orbit had to be properly confirmed before the Moon's motions could be fully comprehended. Even then lunar almanacs could only be produced by day after day carefully measuring the maximum altitude of the Moon and also noting its position to nearby stars whenever possible. Nothing less than a south-facing angle-measuring device fixed firmly in an upright position on a carefully aligned north-south wall would suffice; a mural quadrant. This, coupled with an accurate pendulum clock would provide the exact time and altitude at which the Moon and nearby stars crossed the meridian each night. They rightly concluded an observatory was called for.

Then, two terrible disasters struck London in successive years which understandably diverted their attention.

CHAPTER 4

Outbreaks of bubonic plague, the dreaded Black Death, reached London from continental Europe late in 1664. Everyone blamed the Dutch because there had been an unusually severe outbreak in Amsterdam the previous winter. Indeed Charles II had used this as an excuse to ban all shipping to and from Holland; not that there was much shipping to ban, what with the stringencies of the Navigation Acts. In truth this deadly and mysterious disease was a regular visitor to London; there had been at least seven or eight outbreaks in the previous 100 years.

As if England did not have enough domestic worries, the simmering Anglo-Dutch trade dispute was brought to the boil early the following year when the Royal Navy plundered a Dutch fleet of merchantmen off the Spanish coast near the port of Cadiz. During this action two of the Royal Navy's warships somehow ran aground and had to be written off so the net profit was somewhat smaller than anticipated. The Dutch, despite being escorted by a detachment of their navy had been caught unawares most probably because the two countries were not supposed to be at war.

An English squadron, buoyed by this marginal success then attacked Dutch slave trading posts in West Africa but was seen off by a more powerful contingent of the Dutch navy which was not going to be caught napping twice. War was officially declared once more and the English parliament reluctantly allocated £4 million (£800 million) to the Admiralty to enable them to properly defeat this threat from the windswept flatlands across the North Sea.

It must have been a frightening experience to be in the thick of a 17th century sea battle, where massive floating castles carrying vast amounts of canvas were letting fly at each other with broadsides of 30 or even 40 cannon at a range so close the ships were sometimes touching. These lumbering giants would never have been granted Lloyds' seaworthy certificates and when in fighting trim were dangerously top-heavy. Which is why the main fleets only fought set piece pitched battles in good weather and most certainly not in the winter months.

Whenever possible they sensibly broke off engagements whilst there was still enough light left to plan a safe retreat. Another very good reason for these leviathans to avoid stormy conditions was the danger of literally falling apart because the weight of the guns overstrained the sides of the vessel. 250 tons of bouncing cast iron armaments well above the water line could all too easily rip a wooden ship apart in a heavy sea.

Navigation was still carelessly primitive possibly because these battles were usually fought in shallow water within the confines of the European continental shelf. Contour navigation therefore played a major part in determining position, which is why the charts and advisory notices the Royal Navy made use of, provided more information on depth and sea bed composition than on the actual location of coastlines.

To ascertain position, a lead-weighted line was dropped overboard to take soundings and the mud, shell or gravel that had adhered to the tallow smeared on the lead was carefully examined and hopefully the acquired data could be matched a known position. It is only too easy to imagine the problems navigators faced when a specific location could not be identified in this way. Then the fleet commanders would have to send the manoeuvrable shallower draft scouting frigates ahead to check for sandbars or reefs and search for familiar coastal landmarks.

Later in 1665 at the decisive battle of Lowestoft (chapter 10) the Dutch lost over 4,000 men killed and another 2,000 taken prisoner. The biggest loss of life was caused when *Eendracht,* the flagship of the Dutch fleet blew apart, killing the fleet commander Jacob van Wassenaer and most of the crew; the noise of the explosion was apparently heard over a hundred miles away in London. Although the Dutch were, on paper the overall losers of the 1665 summer fighting season, their opponents failed to deliver the coup de grace and the Dutch made good use of the winter respite to repair and overhaul their badly mauled battle fleet. But the English, even before the battle of Lowestoft had been won were already otherwise pre-occupied.

The plague was spreading through London like wildfire, endangering all those hundreds of thousands of pale-faced creatures living and working in a decidedly unhygienic city whose raw sewage was trundled down to the Thames in huge 'puddings' and loaded onto dung boats each morning. London's most prosperous residents, including the royal court and fellows of the Royal Society, upped sticks and sensibly fled to their alternative residencies in the country. By June 1665 large swathes of the city were silent but for the occasional tolling church bells, the creak of the death carts and the noise of ground-floor doors and windows of contaminated houses being nailed shut; all too often with the dying still inside. This unusual silence may have been why the massive explosion which sank *Eendracht* was heard at such a distance.

Other parts of the country, desperately trying to prevent the spread of the disease, set up road blocks to bar outsiders, in many cases to good effect.

But although by the middle of August deaths in London had reached an appalling 6,000 a week they had mercifully dropped to around 900 by November. The dead were dumped unceremoniously in huge hastily dug pits, at least one of which, to this day has not been built on. Hooke stuck it out until July, interested in discovering the causes, but then caution overruled curiosity and having written a note on the subject to a colleague, he too fled. By March 1666 life in London was back to normal, but there were at least 70,000 less to partake, the city having lost a quarter of its teeming population.

But the war at sea had not been resolved and the English were now in real trouble. The fleet winter repair programme had been chaotic; the plague had spread down-river to the great shipyards of Deptford and Chatham. Manning problems were so acute that press-gangs roamed the streets and prisons were being emptied to find enough men to crew the ships.

Many of these problems could have been resolved had the king been willing to open his purse-strings. Instead sailors were issued with tickets in lieu of wages, leaving entire families starving and many desperate men were forced to exchange their hard won IOU's at well below par to speculators. In direct contrast the Dutch were having little difficulty in obtaining replacements, using the cunning but effective ruse of enlisting the help of English mercenaries who had avoided the clutches of the press gangs, by paying them in hard cash. But worst of all the English Royal Navy had temporarily lost its inspirational fighting royal duke.

The Duke of York had come within inches of suffering a grisly end on board *Royal Charles* during the battle of Lowestoft when a chain shot whipped through his group, killing several and decapitating the son of the Earl of Burlington who had been standing next to him. Charles promptly forbade his brother and heir to the English throne from partaking in the 1666 fighting season, but the navy's loss would be London's gain.

The 'Four days fight' off North Foreland in June 1666 (chapter 10) had gone badly for the English owing to their lack of preparedness, unsolved manning shortages and indecisive commands. But at the 'St. James day fight' in the outer reaches of the Thames estuary two months later, the English managed to get the upper hand and the Dutch lost about 20 ships and 7,000 seamen were killed or wounded. Enough was enough and they sensibly beat a hasty retreat.

An English squadron led by Sir Robert Holmes followed up their resounding victory by chasing the enemy across to the Dutch islands. There they set fire to ships of all shapes and sizes wherever and whenever they could find any. A contingent of marines even landed on the island of

Terschelling and set the town ablaze. This daring action became the talk of London's coffee houses and soon acquired the sobriquet 'Holmes' bonfire'. Coincidentally following his retirement a decade later the swashbuckling Holmes apparently seduced Hooke's 19 year old housekeeper-mistress-niece Grace Hooke.

Then a second disaster struck London. Less than a month after the torching of Terschelling, a fire in the premises of the supplier of double-baked ships' biscuits to the Royal Navy curiously situated in one of the city's sewage exits, Pudding Lane, spread so rapidly that more that 10,000 houses in the City of London burnt to matchwood in three days. Fewer persons lost their lives than one would have expected partly because the plague had already reduced the population dramatically. Many Londoners blamed Dutch arsonists and the Dutch did nothing to deflect these dubious accusations by gleefully referring to the fire as retribution for the recent cowardly 'Holmes' bonfire' attack on their civilians.

It could have been worse. More than 5,000 barrels of gunpowder were stored at the Tower of London which were all safely removed and loaded into barges whilst flames licked at the massive stone ramparts. Even the Tower's Yeoman Warders, (including Edmond Halley's father amongst their number) who rarely got their hands dirty, helped in this dangerous evacuation.

The diarist Samuel Pepys, Clerk of the Acts at the Naval Office at the time, having viewed the fire's rapid spread at close quarters, realised his house would be in danger unless some sort of defence was organised and rushed off to Whitehall Palace to urge his employer, the Duke of York to use troops to help fight the flames. He was dismayed to find the king otherwise engaged (with one of his concubines), whilst about twenty courtiers and other 'dissolute persons' were gambling at a great table, a bank of at least £2,000 in gold coin in front of them, and all seemingly indifferent to the city's fate. Many of the coins carried the Company of Royal Adventurers Trading into Africa elephant emblem.

Whilst the royal party were deciding whether or not to view the progress of the blaze from the Thames aboard the royal barge, the Duke of York was despatched to act on Pepys' report. For the next two days he fearlessly galloped about on his horse organising the blowing up of houses with controlled explosions to create fire breaks. He was in his element and his quick thinking and decisive commands were largely responsible for containing and then finally smothering the flames, although a fortunate change in the wind direction did help.

The Great Fire of London was stopped well short of Whitehall Palace and just short of Pepys' house and his wine, cheese and naval documents which he had sensibly buried in his garden, but only after it had consumed the great city's landmark, St Paul's cathedral. The fire was also brought under control just short of the walls of Gresham College, meeting place of the RS and abode of Hooke. He was quickly co-opted into the city rebuilding project, which was extremely convenient for the penurious Curator of Experiments because a regular income would at last be secured. His lodgings were also secure because the city planners had commandeered Gresham College for their headquarters and he was allowed to keep his apartment.

The RS was not so fortunate and the society's weekly meetings came to an abrupt halt again almost as soon as they had restarted following the conclusion of the Black Death hell, and it was a while before they found alternative accommodation. Unfortunately, because of the fire, the meeting at which Hooke had already announced he would demonstrate his new angle-measuring device (figure 2, chapter 3) was cancelled and the details of this device were to remain unpublished until 1705, two years after his death.

Eventually Hooke and Wren between them designed and rebuilt much of the City of London which was still standing over 250 years later, when Hitler's Luftwaffe made every effort to raze it to the ground again. Over 150 city churches were built or repaired, most supervised by one or the other and often both of these brilliant architects.

Although Hooke played a relatively minor role in the building of the new St. Paul's Cathedral, many of the churches he designed and built with little or no assistance from Wren are amongst the most beautiful of those still standing today. These include St. Edmund the King and Martyr in Lombard Street which is open every day of the week, St. Martin within Ludgate on Ludgate Hill which is open only for Guild services and St. Benet Paul's Wharf on St. Benet's Hill which is now the Welsh Episcopalian church for London.

CHAPTER 5

In the summer of 1667 the Dutch sailed boldly across the North Sea, looted the town of Sheerness, burned, sunk or damaged many of the Royal Navy's warships (including *Royal James*) while they lay at anchor in the Medway river and sailed away with *Royal Charles*, the pride of the English fleet.

Many considered this to be the greatest humiliation England had suffered in 600 years, the loss of the 'royal' vessels especially embarrassing. Pepys, despite being assailed at his office at the Navy Board by unpaid seamen, was so alarmed that the Dutch would invade London that he moved nearly £2000 (£400,000) of his own money out of town; most of which had been obtained by accepting bribes from naval suppliers. The two sides then sensibly called a truce and the (second) Dutch war was concluded.

The finances of the Royal House of Stuart, were, despite the ending of hostilities, in considerable disarray. To cap it all the Company of Royal Adventurers Trading into Africa had not yet yielded as much profit as had been hoped for and by the early 1670's was in desperate financial straits owing over £100,000 to creditors and owning little but the cherished slaving monopoly. Many stockholders were still either unwilling or unable to pay for their stock options and leasing the slave-trading rights to other English companies had not solved the cash-flow problems.

What was needed was a way of re-capitalising the company without having to pay the creditors; and to do that required a considerable amount of guile. The Duke of York would have to convince creditors and stockholders that real progress was being made in maritime navigation. Which is where his personal secretary and fellow holder of stock in CRATA, Matthew Wren was able to assist.

Matthew was Christopher Wren's cousin and good friend and consequently was able to pass on details of Hooke's latest inventions to his master. First, a revolutionary new timekeeper; a watch which, so Hooke claimed, would keep perfect time, even when it was shaken about in a great sea storm. Second, an angle-measuring instrument (figure 2, chapter 3) which was a great advance on the back-staff. Having been assured by Matthew Wren that maritime navigational improvements were well in hand, the Duke of York inflated this assurance further in order to persuade those of his wealthy friends that had not already lost their shirts in the CRATA enterprise to reinvest in the new Phoenix he planned to launch skywards from the ashes of the old company.

An additional attraction for investors was the prospect of also supplying the labour-hungry Portuguese colonies in Brazil with slaves now England had a Portuguese nobleman's daughter as Queen Consort. Suppliers to English and Spanish colonies and now possibly Portuguese as well; the prospects seemed limitless especially now that, according to his experts, trans-Atlantic navigation techniques were about to be improved dramatically.

It was not until the winter of 1671 that investors, including nearly as many of the English aristocracy as before, could be persuaded to underwrite sufficient stock to float the new Royal African Company of England and provide guarantees against the debts of the old CRATA. This guarantee enabled parliament to grant this new trading company exclusive rights to trade and deal in western African gold, silver and slaves, from Cape Blanco in the northwest to the Cape of Good Hope in the south for the strictly limited period of 1000 years. A trading monopoly to an African coastline over 4000 miles in length! Thrown in for good measure; no subjects of the Crown, other than representatives of the new company, were permitted even to visit these lands without permission and the company was empowered to seize the ships and goods of anyone attempting to infringe its monopoly.

The fledgling company was encouraged to establish forts, factories and even plantations along the western African coast and authorised to raise troops and to make war (or peace?) with any heathen nation there. Significantly, as it would later turn out, no mention was made of what the company could actually do with seized ships or goods. Everyone simply assumed it could confiscate without compensation. If the RAC was permitted to make war, surely it need not compensate pirates?

Having successfully survived Cromwell's republican purges, plague and fire, Matthew Wren died in his early forties following wounds he received at the naval battle of Sole Bay in 1672 on board the brand new first-rate *Prince*, which seven Dutch warships managed to isolate from the rest of the English fleet and reduce to a defenceless burning hulk at the start of the third war between the two countries (chapter 10).

Surprisingly the RAC got off to a flying start. The account books indicate that in the 1673-4 season the company exported £20,000 (£4 million) worth of trade goods which included 40 tons of iron bars, a ton of copper bars, brass and pewter ware, woollens and textiles, 300 barrels of gunpowder, 2,200 firearms, 668 dozen knives and a ton of beads. These were bartered for about £8,000 worth of gold and an undisclosed number of slaves which would have cost considerably less than £5 each.

The gold was brought directly back to England and the slaves shipped on across the Atlantic to Barbados, Jamaica and Nevis where the 1,945 survivors were auctioned off like expensive cattle for an average of £17 per head in Barbados and £22½ in Jamaica. Some of the cash generated was used to buy nearly 600 tons sugar which then sold in London for the enormous sum of about £450 per ton; equivalent in modern terms to £90 per kilogram. In all a *net* return on investment of about 70% per annum.

Clearly this was potentially a highly lucrative trade, but losses at sea, dubious book-keeping methods and sticky fingers would ensure that this particular goose would all too often fail to lay the anticipated golden eggs. The gold in question had already been found to be of such purity that each coin minted from this gold from the African Gold Coast Guinea region was worth about 10% more than the gold used by the Royal Mint, which is why the coin known as a 'guinea' had been introduced back in 1664.

The new guineas minted from RAC gold were also impressed with the CRATA elephant or elephant and castle coat of arms which had been adopted by the new company and during the next three decades over half a million of these coins were minted in London, all from RAC African gold. Over the years their value fluctuated in relation to the Unite or Pound (20 shilling) coin; from 21 shillings all the way up to 30 shillings in 1695 at the height of a counterfeit scandal which Isaac Newton and Edmond Halley were in large measure responsible for quashing (chapter 15). But the temptation to divert this valuable commodity at every turn was one of the prime reasons that the official accounts of the RAC would never again appear so healthy.

Charles II guineas depicting the elephant or elephant and castle emblems today fetch up to 10,000 times their face value at auction. The use of the company's coat of arms on the 4 values of gold coins produced by the Mint provided the company with an unusual and long-term advertising outlet, encouraging investors and giving the impression that the company was in large measure responsible for the monetary stock of the nation. One of the earlier examples of inspirational company advertising; but in truth at best RAC gold from West Africa accounted for only 7% of gold coined by the Mint.

Although the French were involved in any number of scientific ventures, the discovery of a reliable method of assessing longitude at sea would of course help them in their continual battle to gain world-wide naval supremacy in the same way that it would aid their age-old adversaries on the other side of the English Channel. The French Royal Academy had been founded in 1666, only three years after the Royal Society had been granted its royal charter. In next to no time Louis XIV had ordered the construction of an observatory in Paris and recruited some of the best scientific brains in Europe to run it. But the over-riding reason he was pouring so much money and effort into solving the puzzle was to prevent his potentially vast profits from the trans-Atlantic slave trade from sinking beneath the waves; French American colonies also relied heavily on slave labour and his trading company was in no better shape than CRATA had been.

At first French progress was rapid. One of Louis' recruits, the Dutch government official Christiaan Huygens, FRS., claimed he was well on the way to perfecting a sea-going pendulum clock although a caustic Hooke doubted it. Jean-Dominique Cassini the Italian astronomer imported to oversee the running of the new Paris Observatory had already improved on Galileo's basic Jovian moons method of establishing longitude on land.

Now a top priority for the observatory was the production of a *reliable* Jovian moons almanac, but from the outset Cassini was hampered by difficulties in plotting Jupiter's elliptical orbit. Nevertheless, armed with the latest estimates, longitude could now be assessed to within a degree or two. But only on nights Jupiter was visible at a reasonable altitude (which was not often), only if the observer could keep his very long telescope steady enough to identify the moment a minute speck appeared to touch the planet's rim and only if local time could be assessed accurately.

Suddenly French research into marine pendulum-driven weight-powered clocks was shelved and Huygens switched his attention to spring-powered balance-wheel regulated watches. He had been forced to switch because his assistant had made the disconcerting discovery that pendulum clocks lost time in lower latitudes. Jean Richter had been sent on a long sea voyage into the tropics in 1670 to test Huygens' latest sea-going pendulum clocks. Seasick almost continuously, Richter nevertheless did his best to look after his delicate cargo. When the ship reached the port of Cayanne in the French South American colony of Guiana, Richter discovered that his precious charges were now running permanently 2½ minutes a day slower than they had done in Paris.

When Richter reported back, Huygens had thrown a fit and blamed the faults on Richter's careless handling of his precious clocks. On a second trip to Cayanne in 1672 the odd behaviour of the pendulum was repeated and Richter and Huygens both realised that for some reason not fully understood, a pendulum would swing more slowly in equatorial latitudes than it did in Paris, Cayenne being only 6° north of the equator. Isaac Newton was later to correctly attribute this to the Earth's equatorial buldge.

The canny Huygens then published *Horologium Oscillatorium* which discussed marine pendulum timekeepers, explaining how they could be used to solve the longitude problem, without mentioning this fatal defect. The French were obviously trying to confuse the opposition, keeping this disastrous set-back under wraps while they worked on an alternative type of timepiece; an alternative that was destined to upset Hooke in spectacular fashion. The results of Richter's unfortunate discovery were only made public seven years later in 1679 and two years after Edmond Halley had 'first' discovered the same effect when he visited St. Helena (chapter 7). Only after he had then visited Paris to discuss these anomalies with Huygens, did Halley realise his discovery would have to be credited to Richter.

At this point in the story a naive Frenchman unwittingly persuaded Charles II to order the building of an English observatory in direct competition to his own country's excellent establishment.

CHAPTER 6

*Note. * in chapters 6 and 7 link to Appendix 2.*

The trail which led to a Frenchman inducing the king of *'England, Scotland, France and Ireland, Defender of the Faith and of the Church of England, and also of Ireland, on earth the Supreme Head'* - to give Charles his full and slightly ambitious title - to build an observatory, is somewhat convoluted.

John Flamsteed was born *'on the 19th day* [o.s.] *of August, at 7h 16m after noon'* [1] in 1646 and wrote a frank and revealing account of his early life 21 years later in what is best described as a personal diary apparently written, unlike those of Hooke, with a degree of hindsight. A story of success through adversity with the help of God, although considering all the ills and misfortunes which befell him he could have been forgiven if his faith had wavered. Born in the village of Denby seven miles north of Derby in central England, his mother died soon after giving birth to his sister when John was only three years old and he was brought up by his father who left him to his own devices for long periods. Like many of his generation, the mysteries of the heavens fascinated him.

He was by his own account a sickly child, and his health deteriorated further after swimming in a river when he was 14; in his own words, *'it pleased God to inflict a weakness in my knees and joints upon me.'* [1] Sadly he never recovered and over the years his infliction worsened despite leaving his village to visit a faith healer in Ireland at the age of precisely *'19yrs.,6days,11hrs',* [2] a second statement so precise it would at first sight imply that someone in the Flamsteed household kept a diary stretching back to the date of his birth to which he could later refer. But on close examination these precise timings are not what they seem and are merely a budding astronomer's way of saying '(my mother told me) I came into this life just as the Sun was setting' and 'I remember leaving for Ireland an hour after sunrise six days after my birthday'. The Sun set at the precise minute at the location, date and time Flamsteed provides for his birth and rose one hour prior to '19ys.,6days,11hrs' later.

Remarkable historical solar computations made by a young man in 1667, which in those days only another very good astronomer would recognise as such. The Irish faith healer's administrations had no effect but on returning safely to Denby, now so ill that he could not attend church as usual, he wrote *'For God's providences in this journey His name be praised. Amen, Amen'.* [3]

In between bouts of sickness in 1669, he predicted *'some remarkable eclipses of fixed stars by the moon, that would happen in the year 1670'* [4] and submitted these details to the Royal Society signing his letter merely 'J.F.'. This brought him to the notice of Sir Jonas Moore, a keen amateur astronomer and the wealthy Surveyor General of the English Ordnance Office and it was through him that he obtained his first telescope and the micrometer fitting for it.

He immediately began observations, persisting doggedly despite suffering, like Hooke, from regular migraine attacks. It was at about this time that he first became aware of Isaac Newton's activities, being fascinated by the thought that the white light from the Sun and stars might actually be made up of a whole range of different colours, some of which he had observed around the edge of his telescope's lenses and which had exacerbated his bouts of migraine. In March 1672 he read about Newton's exciting *'new-contrived telescope'* [5] (chapter 8).

On 8th June 1674 Flamsteed paid a visit to Jesus College Cambridge and was granted a Master of Arts degree by royal mandate which permitted him to avoid the normal residency requirements because he intended to take holy orders. He did not however seek religious employment because *'...the Good Providence of God, that had designed me for another station, ordered it otherwise...'* [6] ; Moore had just offered to fit up a house as an observatory and to put the new Cambridge graduate in charge. Flamsteed clearly saw this offer, not as a slice of good fortune but as preordained. If God in His infinite wisdom had commanded him to be an astronomer not a priest; so be it. And it would not take the young man long to convince himself why he had been chosen; the French couple Louise de Keroualle and the Sieur de St. Pierre unwittingly saw to that.

Louise was a French maid of honour to the sister of Charles II, Henrietta, Duchess of Orleans who had been one of the original investors in the Company of Royal Adventurers Trading into Africa. Louise had been brought to England by Henrietta in 1670 and brother Charles, having already worked his way through any number of mistresses, fell for her baby-faced charms.

Installed in her own apartments in Whitehall Palace, Louise became the king's latest official mistress with a rumoured allowance of £40,000 (£8 million) a year from the royal coffers, which goes some way to explaining why these were never very full. Nell Gwynne, one-time orange seller and now one of the king's other current mistresses was not amused. In the hearing of the king she would often refer to Louise as 'dearest squintabella' (owing to the slight cast in Louise's eye), or 'the weeping willow' (because of her infuriating habit of silently letting tears fall in order to get her way). Oddly Charles found Nell's comments amusing, but nevertheless, was not dissuaded from creating Louise the Duchess of Portsmouth and speedily siring a son by her.

St. Pierre was a marginally gifted French astronomer who, not having done his homework, was convinced he had solved the riddle of assessing longitude at sea. Excitedly he had taken his case to Louis XIV's experts who listened politely, explained the drawbacks to his idea and then threw him out. Upset at this lack of respect and convinced his case was sound, the angry Frenchman crossed the Channel and persuaded Louise that she could make a profit by informing her lover that he had an important secret to sell.

The king, pre-occupied with mistresses and spaniels, in turn passed the case to his brother and in December 1674 a Royal Commission was established to examine the feasibility of this longitude proposal and another, a geomagnetic idea, under the direction of Moore and Lord William Brouncker, President of the RS and member of the Navy Board.

Other members of the commission included a number of *'ingenious gentlemen about the town and Court',* [7] and RS fellows Seth Ward (the Bishop of Salisbury), the recently knighted Christopher Wren and Hooke. Hooke in turn and to his everlasting regret, agreed to Moore's suggestion that they co-opt the 28 year old Flamsteed.

The geomagnetic submission to the commission was, according to Hooke based on only two local observations and he rightly branded the unfortunate proposer *'an incompetent amateur'.* [8]

But when they got down to the serious business of dealing with 'dearest squintabella's' French friend, it soon became apparent that St. Pierre was trying to sell them the notion that longitude could be determined by the basic lunar 'heavenly clock' method, centuries after it had first been suggested.

Like their French counterparts, the examining group of experts also pointed out the very serious flaws in his presentation and then showed St. Pierre the door; but the stubborn man did not go willingly. In protesting that he had been badly treated he ran foul of Flamsteed.

St. Pierre demanded evidence that his (and everyone else's) lunar data was woefully inadequate for the task. Flamsteed produced the evidence and St. Pierre promptly claimed this had been fabricated. So Flamsteed proved his point in a report to the commissioners [9] (appendix 1), thus sealing the irate St. Pierre's fate.

Having rid themselves of the Frenchman thanks to Flamsteed's expert intervention, the commission members were quick to see the brilliant opportunities St. Pierre's flawed proposition had presented. First they agreed that the lunar-distance method of determining longitude at sea was indeed the one to go for. They then made excellent use of their new lunar expert Flamsteed's report on St. Pierre, which had slightly exaggerated the flaws and glossed over the monumental difficulties which would be encountered in attempting to correct these inaccuracies. The plan was to persuade the king to authorise the construction of an entirely new observatory, not to put him off the whole idea.

Exactly who was responsible for compiling the commission's report is impossible to determine, but certainly Wren, Hooke and Flamsteed must all have been well aware of their misleading advice. No record of the actual report appears to have survived; possibly because it may have been delivered verbally to Charles by his childhood friend Wren; if so, a wise decision bearing in mind St. Pierre was a protégé of the king's official mistress.

Flamsteed's version of events was that, in dismissing St. Pierre's suggestion, he reported to the commission that *'the Catalogue of the fixed stars made by Tycho Brahe....and now used, was both erroneous and incomplete: - that the best tables of the moon's motions....erred sometimes above 20 minutes* [of arc]*; which would sometimes cause a fault of 15 degrees, or 300 leagues in the determination of the longitude by it.'* [10] (see appendix 1 for the full quotation).

Indeed the Brahe catalogue was erroneous and incomplete and perhaps some of the tables did occasionally err by 20 minutes of arc, but the handful of astronomers currently working on recording stellar positions and the Moon's passage through them were certainly not in error by that much. That's as maybe, but to imply that a 20' angular error would result in a 15° error in longitude was not correct unless the navigator was foolishly trying to measure the angular distance between the Moon and an entirely inappropriate star; when the errors could be much greater still.

Regardless of any report massaging, the commission rightly recommended that every effort should be made to make the lunar distance method workable. Sensibly no one mentioned the inconvenient fact that it would take at least 18 years before anyone could hope to produce data of sufficient accuracy to enable mariners to determine their longitude on the high seas by this method; they simply recommended that an observatory be built, designed by Wren, supervised by Hooke, run by Flamsteed and, dare they suggest, financed by the king?

The audience, according to Flamsteed's précis, went as follows:- *'his Majesty would give a great and altogether necessary encouragement to our navigation and commerce (the strength and wealth of our nation) if he would cause an Observatory to be built, furnished with proper instruments, and persons skilful in mathematics, especially astronomy, to be employed in it, to take new observations of the heavens, both of the fixed stars and the planets, in order to correct their places and motions, the moon's especially; that so no help might be wanting to our sailors for correcting their sea charts, or finding the places of their ships at sea. Hereupon his Majesty was pleased to order an Observatory to be built in Greenwich Park: Mr Flamsteed was appointed to the work'* [10] (appendix 1).

Note the clear emphasis on maritime navigation and lunar observations from the very outset. Hooke and Wren, despite their heavy work load had somehow conjured up planning permission and finance for an observatory; the nation's longitude quest was up and running thanks to a Frenchman.

Acting on Wren's advice the king gave orders for an observatory to be built to Wren's design well away from the polluted atmosphere of London in leafy Greenwich Royal Park. In order to save time and money the site of an existing building on the highest point in the park, the remains of Duke Humphrey's Tower, was selected, which incidentally would provide stunning views across the Thames to the north. An overworked Hooke undertook to act as Wren's site manager and supervise the construction which commenced in July 1675.

As if Hooke did not have enough on his plate, he was also at this time entangled in a dispute with Louis XIV's pendulum clock specialist Huygens. In January 1675, one of the Secretaries of the RS, Henry Oldenburg had read out a letter from Huygens to an assembly of fellows. Their Dutch corresponding member was advising them that he had invented a new watch but withheld the details by concealing them in a Latin transposition cipher. The English 60 letter version would have read...

aaaaaabccccdeeeeeeeeffhhhhiiiiilllmnnnoooooprrrrsssttttttvx

A computer programmed with the likely text matter could probably unravel this in a few seconds, but in the 17th century it was unbreakable. Mind you, a clever person could hedge his bets should further research warrant it, by producing a solution having an alternative meaning (chapter 24). Even the mere transposing of words might be sufficient.

A month later a second communication provided the key which, translated from the original Latin revealed the details, such as they were. *'The axis of the movable circle is attached to the centre of an iron spiral.'* Even if no one else understood the import of this claim, Hooke certainly did. It meant that Huygens had abandoned the swing of a pendulum in favour of a spiral spring's oscillations as the driving force for a new marine watch, still incidentally without revealing why the pendulum was being discarded.

Hooke loudly protested to the RS and to his coffee house audiences that he had invented and manufactured watches in the same manner more than 10 years earlier, *and* had revealed details in a Gresham College lecture in 1664 *and* shown one to the society in 1666. Whilst his coffee house supporters applauded loudly, fellows remained unimpressed, suggesting cattily that Hooke's spring watch had not worked properly. A fair point but no one knew if Huygens' watch worked either.

To add insult to injury (bearing in mind Hooke's own earlier patent problems (chapter 3)), within weeks Huygens, who could not hold an English Royal patent because he was a foreign national, had offered Oldenburg the patent if Charles II could be persuaded to grant it.

Hooke was livid, even more so when he discovered that his friend Sir Robert Moray had written to Huygens back in 1665 to advise him of his spring inventions. Nothing wrong in advising overseas members of that which had been revealed in a semi-public lecture of course, especially as Moray had probably only written intending to establish Hooke's priority. But Huygens' defensive criticism of Hooke's watch *('Mr Hooke speaks a little too confidently of this invention of the Longitude, as in several other things')* only served to spur Hooke on to louder protests.

At this point the incensed inventor was approached confidentially by one of the king's tax collectors, Sir James Shaen. Could Hooke really construct a spring-regulated marine timekeeper as he was claiming? Ever the optimist, Hooke implied that he certainly could so long as the reward was sufficiently great. Shaen offered him a lump sum of £1,000 (£200,000) or alternatively £150 (£30,000) per annum for life. Hooke plumped for the annuity [11] and Oldenburg dropped the Huygens patent application like the hot potato it was.

Hooke then frantically attempted to improve on his own design to make it truly workable, but to cut a very long story short, he never did succeed. This was partly because he was trying, as usual, to wear several hats at the same time, and partly because neither he, nor anyone else could properly work out how to compensate for the expansion and contraction of metals and spring flexibility caused by temperature changes.

Meanwhile across the Channel the French decided that the key to the longitude at sea problem rested on the ability to manufacture a reliable sea-going chronometer. To this end every encouragement was given their

Dutch expert Huygens who devoted the rest of his life (he died in 1695) trying unsuccessfully and in secret to produce a watch that could keep good time at sea. Thus did the French and quite possibly the Dutch, having strongly backed the clockwork solution, severely handicap their own chances in the longitude race.

Although it would be many years before Newton publicly expressed his thoughts somewhat forcefully (chapter 26), it had become painfully obvious as early as 1680 that achieving a longitudinal fix by watch-work *alone* was logically impossible. One also required an accurate angle-measuring device capable of being used in rough weather and of being automatically corrected whilst on the high seas if damaged. Without the invention of such a device the watches of Hooke or Huygens would be of marginal use even if either inventor had succeeded in manufacturing a reliable version.

Hooke then discovered he was supervising the construction of the observatory on a shoe-string budget! Although the king had implied he would finance the venture, Charles actually contributed little but name plus some second-hand building bricks taken from an old fort and wood, iron and lead from a demolished gatehouse. So most of the hard cash had to come via the ingenuity of Moore and his Ordnance Office. He auctioned off the rotten gunpowder that had been left over from the second Dutch war years earlier; powder salvaged from the Tower of London at the time of the Great Fire. Being in charge of ordnance, disposal of this time-expired surplus seems to have presented no real difficulties.

But the sum raised fell short of expectations which suggest that some of the purchasers knew how old the powder really was and Moore eventually had to dip into his own pocket to enable the construction to be completed.

Using the old foundations which had been sensibly well laid on rock, was the speediest and cheapest method. But this meant that none of the walls would be aligned properly on a north-south axis. Wren and Hooke were of course both well aware of the importance of a north-south wall for fitting the essential mural quadrant (which Hooke hoped to sell to Moore) in order to measure meridian angles; the point at which heavenly bodies reach their zeniths. They would just have to construct this wall in the grounds, a wall which unfortunately would not be built on the firmest of foundations. Unfortunately this latest penny-pinching was to cost Flamsteed dearly in years to come [12].

Wearing his surveyor's hat Hooke was responsible for the rapid completion of the observatory, which was finished within a year and included lodging rooms for the king's Astronomical Observer and an assistant. Flamsteed meanwhile carried on his observations from the Queen's house at the bottom of the hill in Greenwich Park.

The grand opening of the Royal Observatory had been stage-managed to coincide with a partial solar eclipse on the 11th June 1676 * and the king's presence had been eagerly anticipated by all involved in the project. He failed to put in an appearance, probably because at 8 o'clock in the morning he had only just fallen asleep and, unlike ancient eclipse rituals, there would be no virgins on display. In the event, few of the instruments had arrived because of the speed at which the building had been completed and those worthies that did get up early witnessed a somewhat unspectacular event when the Moon managed to black out one third of the Sun.

At least the construction of the observatory enabled the English to announce a new prime meridian and that they no longer recognised Paris as holder of that honour. This unfortunately meant that the two countries would henceforth have to convert each other's astronomical data before being able to make use of it.

Flamsteed was in business with an annual salary of £95 after tax (considerably less than Hooke had been offered for his watch), and a very specific job description. The key royal command was that he was to devote himself *'with the most exact care and Diligence to rectifying the Tables of the Motions of the Heavens, and the places of the fixed Stars, so as to find out the so-much desired Longitude at Sea, for perfecting the art of Navigation.'* [13] A straightforward regal order; solve the longitude at sea problem by astro-observation, an instruction which in 1700 and again in 1707 Flamsteed had no difficulty later noting in his diaries. Almost as an afterthought the appointment warrant neatly placed the responsibility for providing Flamsteed's salary on the shoulders of the Royal Ordnance Office.

So the privately owned Royal African Company had somehow obtained a brand new national *maritime* navigation research station, which with luck would save their stockholders a fortune without them having to pay so much as a penny piece for it. Significantly the RS also had a financial interest in the success of the observatory because they held £200 (£40,000) in RAC stock at this time. All the Duke of York now had to do was to ensure that the scientist whom the Ordnance Office was paying, produced the results his company needed. Unfortunately he made no allowances for divine intervention.

Flamsteed had endorsed the Royal Commission's report which had sensibly and rightly drawn attention to the defects in current knowledge of the lunar orbit and positions of major stars. He had also implied that these deficiencies could be easily rectified if only an observatory were to be built. Now he had to do the rectifying by spending long hours in an unheated outhouse with his eye glued to the eyepiece of a telescope whilst suffering from migraine.

Noting the exact height, diameter and time the Moon crossed the meridian every 25 hours or so and obtaining similar information on major stars with emphasis on those on or near the ecliptic should have proved relatively simple, if tedious, once his instruments were in place. But he knew only too well that it was going to be a long job; 18 years and then some! Only then could he even begin to consider publishing the lunar almanac which would fulfil his contract of employment; 1694 at the very earliest.

Flamsteed was infuriated to discover that he was expected to fund the cost of any instruments from his salary over and above any that Moore and the Ordnance Office were supplying to start the venture. This cheapskate treatment was to discolour Flamsteed's opinion of officialdom for all time and he had to set about his task as best he could with limited instrumentation. Moore paid for two 13ft. pendulum clocks made by Thomas Tompion and a 7ft. radius sextant set on an equatorial mount made by Moore's Tower smiths and fitted out by Tompion [14]. The sextant was equipped with a telescope which had Flamsteed's own reticule sight within the tube (an insert on which cross hairs were mounted) which permitted very accurate measurements. But Flamsteed himself had to provide a 52ft. telescope, two smaller ones and *'an indifferent small quadrant'*.[15]

The essential south-facing wall-mounted 10ft. mural quadrant [16], which Hooke had eventually succeeded in selling to Moore was delivered a month after the official opening and it was at this point in the story that Hooke and Flamsteed became bitter enemies.

The overly optimistic and overworked Hooke had supplied a vital but defective piece of equipment; the special measuring attachment fitted to the mural quadrant was a quirky and dangerous untested invention. As Flamsteed was to write; *'I have often tried to make Mr Hooke's wall-quadrant give me altitudes, and it has as often deceived me, and lost its rectification. I tore my hands by it, and had like to have deprived Cuthbert* [an assistant] *of his fingers.'* [17]

Flamsteed should have insisted on Hooke repairing or replacing the defective item; instead he tried unsuccessfully to fix it himself. Within six months it had been removed from the wall and Flamsteed replaced it temporarily with his *'indifferent small quadrant'* and shortly after with a larger mural quadrant loaned(together with three other items) by the RS [18].

But Flamsteed was never really any happier with the RS instrument than he had been with the useless Hooke model, or so he claimed.

Exactly when Flamsteed decided on precisely what God had in mind for him is unclear. Possibly ever since he had become interested in astronomy as a young lad, or possibly only after he realised that his true terms of employment as the Royal Astronomical Observer were beyond the scope of his instruments. Or possibly because he suddenly received that message from on high after graduating or when the faith healer failed him or perhaps it all came to him out of the blue after a bad day at the office. Whatever, within a couple of years of his appointment he had already clearly come to the conclusion that the station his Maker had allotted for him was the formidable but attainable task of charting the position of every visible star in the heavens and then publishing maps of the various constellations to enable all true believers to marvel at His works.

The king's orders would be carried out only as and when conditions permitted. But he would have to be careful not to make his priority obvious; he could lose his job and that would be self-defeating. Mapping the heavens became his hidden agenda and his hair shirt.

Which is why Flamsteed came to dislike mural quadrants in general; the mere fact that they were quadrants. They were fixed firmly to walls, aligned with extreme care and could measure angles up to 90° but no more. Most of his observing time should have been allocated to plotting the meridian positions and movements primarily of the Moon, but also of the planets, Sun and the brighter stars close to the Sun's path (the ecliptic) which were confined to a specific *southerly* aspect of the Greenwich sky.

These were the objects which, when properly measured with a mural quadrant would solve the longitude puzzle, not stars elsewhere in the heavens. But such an instrument could not measure the positions of stars in the northern portion of the sky and enable him to complete his star maps. Many years later, when at last he could afford one, he fitted a quality mural arc to the 'Hooke' wall which could measure 140° and enable him to fulfil his mission, or so he fondly believed. But somewhere along the line God had let his disciple down; adding that extra 50° was a biggest mistake of Flamsteed's painful life (chapter 11).

Meanwhile in the absence of a reliable fixed quadrant, Flamsteed could argue with some justification that there was no point in wasting too much time on taking unreliable measurements relating to his official remit when there was so much other astronomical work to do. So he happily set about plotting the exact position of every star in God's Universe visible from Greenwich with his Tompion 7ft radius iron swivel sextant (he eventually mapped some 3000), as well as carrying out his official work at the same time as best he could. But, unlike a fixed mural arc, the swivel sextant could only measure the distance between heavenly objects; it could not properly register each star's exact position in the sky.

When others began requesting lunar data, Flamsteed's inadequate funding allowed him to argue that any information collected at the observatory was his personal intellectual property, thus masking for many years the paucity of such data. By implication, what he studied at Greenwich was not the business of others. But Greenwich Observatory was not Flamsteed's private observatory and the argument was invalid, as Newton would one day angrily point out to him.

(1) Francis Baily, *An Account of the Revd. John Flamsteed* (London, 1835/1966), pp.7-8.
(2) Ibid., p.13. (3) Ibid., p.20. (4) Ibid., p.28. (5) Ibid., p.29. (6) Ibid., p.36. (7) Ibid., p.37.
(8) Stephen Inwood, *The Man Who Knew Too Much* (London, 2002), p.193.
(9) F.Baily. p.37-8 & 187. (10) Ibid., p.187.
(11) Thomas Birch, *The History of the Royal Society of London* (London 1756/7), Vol. 3, p.191.
(12) F.Baily. p.40 site plan of observatory.(13) Ibid., pp.111-2 (14) Ibid., p.41. (15) Ibid., p.45. (16) Ibid., p.43. (17) Ibid., p.118. (18) Ibid., p.45.

CHAPTER 7

*Note. * in chapters 6 and 7 link to Appendix 2.*

Edmond Halley, in many ways the hero of this story, was born in the late autumn of 1656 within the sound of the bells of St. Mary-le-Bow church; which by tradition made him a cockney native of London town. Ten years later the 600-year-old church was burned to the ground during the Fire of London and was one of the first to be rebuilt by Christopher Wren. In that same disastrous fire St. Paul's School, where young Halley was being introduced to the mysteries of basic mathematics was also destroyed and until its re-opening five years later he was taught privately.

Whoever was responsible did a wonderful job, because by the time he was admitted to Queen's College, Oxford at the age of 16 in 1673 (by which time his mother Anne had died) he had already acquired an enviable reputation as a mathematician and budding theoretical navigator who had taken many astronomical and magnetic measurements.

Even before graduating, Halley had met Christopher Wren and Hooke at Garraway's coffee house, at a time when the two architects were still deeply involved in the planning stages of London's reconstruction following the Great Fire. By all accounts Hooke had been impressed by the youngster's intelligent questions and rapid grasp of the highly technical answers; high praise indeed.

Halley was also about to meet the 28-year-old Reverend John Flamsteed following on from a letter he sent to the newly appointed Royal Astronomical Observer in March 1675. Flamsteed had invited Halley to observe a partial eclipse of the Sun predicted to occur as dawn broke on 23rd June at about 4 o'clock in the morning *. Unfortunately the sky was cloudy and their first collaborative effort was foiled.

Significantly, as it was to turn out, Halley's letter had advised Flamsteed of his observation of a lunar eclipse on 11th January of that year *. He mentioned that he and Thomas Streete had timed the exact moment when the whole disc of the Moon was immersed in the Earth's shadow, the moment of emergence and the ending of the eclipse. Halley even recorded the immersion and emersion (occultation) of a 6th magnitude star * that occurred during the eclipse; a star that would one day be designated 85 Geminorum in Flamsteed's great star catalogue.

Halley and Streete had linked these observations with the position of the star Pollux which, together with further observations of other bright stars later in the month, convinced Halley that the famous star catalogue of Tycho Brahe contained errors [1]. He was right, and it is an evidential point that Flamsteed was informed of 18-year-old Halley's independent discovery, which matched his own, a discovery Flamsteed had recently used to encourage the king to sponsor the construction of Greenwich Observatory. Halley had also mentioned that he was *'reasonably well provided in instruments which I can confide to one minute without error by means of telescopicall sights and a skrew for the subdivision.'* Which was possibly better than Flamsteed could manage at the time, given he did not take up residence in his new observatory until 20th July 1676.

Halley, clearly now fascinated by the navigational possibilities which lunar occultations (the moon passing in front of a star) and appulses (the moon passing close to a star) presented in relation to the longitude quest, made a number of such observations from Oxford during the summer of 1676. A study, which had begun with 85 Geminorum in January 1675, would soon lead to Halley possessing more first-hand information on the subject than Flamsteed, the man now employed specifically for this task. Because Halley kept Flamsteed informed of his activities, the latter could not fail to perceive the young man as a threat to the achievement of his hidden agenda.

Halley's blossoming talent was rewarded early the following year with a trip to St. Helena, an island in the South Atlantic where the tiny population derived mainly from some 30 Londoners who had been left homeless by the Great Fire and had emigrated to the remote outpost 10 years earlier. The cost of the expedition was paid for by his father and sponsored by an impressive gaggle of influential personages headed by Moore and Secretary of State Sir Joseph Williamson. His king personally instructed the East India Company to ship him there and he had the support of the Astronomer Royal [2] as Flamsteed was now calling himself.

Although Halley had been sent to observe the transit of Mercury on 7th November 1677, an event that would take all of 5 hours and 13 minutes, he and a friend actually spent an interesting year on the island.

He established the island's latitude accurately with the aid of a back-staff fitted with a Flamsteed lens [3] (see also chapter 13) and checked the longitudinal position as best he could with the aid of a partial lunar eclipse early on the morning of 17th May 1677 *. The 6°40' W (of Greenwich) result was a full degree in error; had he been able to obtain a Jupiter fix, which was denied him because of the almost constant cloud cover and strong winds which shook his big 24ft. telescope, he would surely have not made such a mistake. An error that was to cause him problems 23 years later. He also plotted the positions of a large number of southern hemisphere stars with a heavy 5½ft. radius brass sextant, tested his big pendulum clock (which had been shipped in pieces) for accuracy and discovered it swung more slowly [3] (chapter 5).

How did Halley discover his pendulum clock was swinging slowly? Clocks and watches could (and can) be checked for accuracy by lining up a star against a mark on a window edge and a nearby chimney, church steeple or even a post hammered into the ground. The star should disappear from view 3 minutes and 54 seconds earlier the following night because one complete spin of the Earth takes 23 hours, 56 minutes and 6 seconds, not a full 24 hours.

Whilst Halley was enjoying life in the southern seas, a depressed and frustrated Flamsteed got wind of a chance to better himself; his short-lived royal appointment, severely under-funded, was not living up to expectations.

The prestigious Savilian chair of Astronomy at Oxford University was about to be vacated by the incumbent, Edward Bernard who apparently was intending to swap a life of astronomy and mathematics for a theological one. Flamsteed, realising that here was a chance to make real progress with his star-mapping programme in relative freedom, wrote to Bernard asking for his advice on how to apply for the post, pointing out that he was not acquainted with the formalities, being a Cambridge man [4].

Flamsteed's intention to leave Greenwich must have reached the ears of prominent opponents early on and certain factions within the Royal Society decided to try to get the society's loaned instruments returned to hasten his resignation given that the Astronomer Royal was spending most of his time complaining and still had not submitted any results for publication. The subterfuge was doomed to failure because of Moore's influential patronage and the instruments remained at Greenwich [5].

Bernard advised Flamsteed that because he was not an Oxford graduate, Halley would be the most likely choice if he were to resign the chair. Bernard then received the following reply *'...my inclinations are for an employment that may render me more useful to the world, and promote more glory to my Maker, which (as you well intimate) is the sole end of our lives, and to which I would divert all my labors'* thus revealing Flamsteed's real astronomical intentions. He then carefully added the following gratuitous comments; *'Last week I received a letter from Mr Halley, who tells me that, if the clouds (which are more frequent than he expected) prevent him not, he hopes to be home by August next. But I am apt to think he will make it Christmas ere he returns. He is very ingenious, as I found when he talked with me: and his friends being wealthy, you may expect that advantage by a resignation to him, which is scarce in my power to afford you.'* [6] In other words, having appealed to a fellow theologian, he added that although Bernard might think he would benefit financially from promoting Halley's cause, Halley would not be home in time to be considered.

But Halley had actually written to both Flamsteed and Moore telling them that he planned to leave for home in February or March at the latest [7] with no mention of *'hoping to be home by August'* and was in fact back in London by the middle of May as planned and a full six months earlier than Flamsteed had implied. This was the first of a series of half-truths, lies and libels to be directed at Edmond Halley by John Flamsteed, every one doubtless fully justified in the latter's own mind as being in the interests of a greater cause. In this instance Flamsteed's efforts were to no avail; Bernard did not resign the post until 13 years later in 1691 by which time Halley was a well-qualified candidate and Flamsteed would again interfere (chapter 11).

Immediately on his return from the South Atlantic, Hooke was kind enough show the Halley star map of the southern skies to the RS who immediately arranged for it to be published [8]. Moore and Hooke then suggested that Flamsteed might consider publishing his own data in like manner; Moore, his patron going so far as to suggest in July 1678 that unless he published an account of progress being made at Greenwich he would consider having Flamsteed's salary stopped [9].

Stung by this very real threat, and not having the lunar-linked data to publish, Flamsteed employed his backhanded methods again by informing Moore of the considerable amount of work he had already done relating to stars and planets adjacent to the ecliptic, of his ill health and of the real problems he faced in the running of the under-funded observatory. But above all until he had sorted out the errors in Tycho Brahe's data there was little point in publishing; a stalling tactic which Moore was in no position to dispute.

To underline his contention Flamsteed told Moore that Halley's star map was as good as could be expected, given the limited time at the young man's disposal, and *'I would not therefore have you understand anything I shall here write concerning his works to his disadvantage'.* Then promptly went on to do exactly that by claiming that in tying his star positions into the old Tychonic framework, Halley's data were suspect [10].

This was the second of Flamsteed's untruths directed at Halley, on this occasion as a means of diverting attention away from his own hidden agenda. This particular attack on Halley's integrity presumably paid dividends because Moore's threat was never implemented.

Others had been much impressed with the young man's results; his king recommended to Oxford University that they grant him a Master of Arts degree [11] and Moore successfully proposed Halley to the RS for a fellowship. In August of that same eventful year Flamsteed travelled back to the family home in Denby having been smitten with *'a dangerous illness'* of some undefined nature [12] and, at Moore's suggestion Halley deputised for him for the rest of the summer of 1678 carefully registering several lunar and Jupiter positions.

On the evening of 24th September Halley, with the help of a friend observed an unusual lunar event; rho^2 Sagittari was occulted by the Moon whilst rho^1 was very close *.

When Flamsteed returned to work they, together with Peter Perkins made careful timed observations of a total lunar eclipse on the evening of 29th October * and John Colson made observations two miles to the west; Cassini did likewise in Paris. As a result of this joint co-operation the longitudinal difference separating the two capitals was fairly accurately determined; Paris was 2°20'E of Greenwich if you were English, or Greenwich was 2°20'W of Paris if you were French! Either way everyone was right for once [13].

Halley, then at a loose end, was asked by Hooke if he would consider a trip to Danzig (Gdansk) to verify the recent claims of the famous astronomer Johann Hevelius.

Hevelius had stated that he could take stellar measurements over open sights that were as accurate as those using a telescope fitted with a reticule. Not that Hevelius was lacking such equipment; he was an old man still possessing exceptional eyesight who on occasions found open sight observations less fatiguing.

Egged on by Hooke, some RS fellows had expressed incredulity and as Halley was familiar with both methods, everyone had thought it would be a good idea to send him to check.

Halley arrived in the Baltic trading port with a well-deserved reputation that had preceded him, having sensibly kept the great man informed of his work on St. Helena. Just in case it should be thought that Halley had come to check Hevelius's credentials, the RS had written to tell him that Halley was intent on visiting the 'Prince of Astronomers'.

The Hevelius observatory was housed in an old brewery that doubled as a family home and at the time of Halley's visit Johann was 68 and his second wife Elizabeth 33; the couple had four children, the youngest eight years old. Halley was given a friendly welcome and began observations as soon as he arrived on 26th May 1679.

Together with Johann, Elizabeth and as many as four other assistants he spent the following two months checking the accuracy of the Hevelius open sight observations. However, he also took every opportunity to check for lunar occultations and appulses using the observatory's 12ft. telescope. Unfortunately, there were not many such events to observe during those short summer nights on the Baltic coast.

Apart from accurately timing the eclipse of Jupiter by a crescent Moon on 5th June * [14] which would enable the difference in longitude between Danzig and Greenwich to be precisely determined, he observed the occultation of two stars by the Moon on 24th of June * and a useful appulse the following night *, all observations which would provide very important data on parallax and angular differences when compared to matching observations made by Flamsteed at Greenwich.

The two occultations of the 24th were especially important because it involved the stars rho^1 & rho^2 Sagittari, the careful measurements of which Halley had taken in similar circumstances the previous September at Greenwich.

Halley did not spend all his time studying the heavens. In a letter to Flamsteed he noted somewhat ambiguously that he had been *'wholly taken up with the Curiosityes of this Place'* and pleaded with him *'to be more than ordinarye intent upon the occultations and appulses dureing my absence.'* [15] Clearly this hot-blooded young man had been temporarily diverted by certain unspecified curiosities, but not sufficiently to forget to remind Flamsteed of his intention to compare notes, especially those on rho^1 and rho^2 Sagittari on his return to London.

Differences in priorities were obviously surfacing and it was Halley who was focussed on the Astronomer Royal's regal 'navigation' remit whilst Flamsteed himself was faced with the daunting task of fending him off whilst at the same time trying to complete his star-mapping programme. At least he did not have to compete with an opponent who was installed in the Savilian chair of Astronomy at Oxford.

Something untoward happened on the 17th July that prompted Halley to write a slightly exaggerated testimonial on the open sights method when used personally by Johann Hevelius and to set out on an unscheduled return journey to London the following morning. In the light of what was to follow, Flamsteed's subsequent rumoured allegations of impropriety were possibly merited and it is easy to imagine that Elizabeth would have found the tall and handsome young Englishman sexually attractive in the circumstances. It is equally easy to imagine that Halley, if caught 'flagrante delicto' might prudently decamp.

Two month later the entire converted brewery in Danzig mysteriously burned to the ground and Hevelius apparently died in the flames. Halley then wrote to a close friend of the Hevelius family offering his condolences to Elizabeth and assuring her he was sending the silk dress she had asked him for before he had left Danzig in such a hurry. It seems odd that he did not write to Elizabeth directly on this matter, but perhaps he was not prepared to become too closely involved with a young widow with four children. Halley was surprised to receive a letter two years later from Johann Hevelius enclosing money for the dress; he had not, after all died in the fire!

When Halley visited Flamsteed after his early return to London to compare notes on the timing of the Jupiter eclipse and the Sagittari events, even he must have noticed a slight cooling in their relationship, a relationship that was in reality akin to that of a venomous cobra with defective eyesight and an amiable young mongoose.

Had Halley, whilst running the observatory the previous summer discovered the surprising lack of lunar data; that Flamsteed had not been performing his functions in accordance with his remit? He would have to have been as poor sighted as that cobra not to have done so. Did Flamsteed think Halley had passed his discovery on to others? Probably, and in the light of what was about to happen, the Reverend's fears must have seemed well-founded.

Two weeks after Halley's return to London in August 1679, Flamsteed's patron Sir Jonas Moore died unexpectedly and again the RS concluded that Flamsteed would have to be replaced if the longitude at sea puzzle was ever going to be solved.

The problem they still faced was that the Astronomer Royal's appointment was for life and whilst Flamsteed was in continual poor health, he had made a miraculous recovery from last summer's grave illness and now gave no indication of actually dying in the near future. But now Moore was out of the way they might be able to make his position so uncomfortable that he would have no choice but to resign.

Hooke, Wren and Sir John Hoskins now managed to remove the four instruments that had been on loan from the RS including the observatory's only working mural quadrant and thus effectively prevented Flamsteed from taking those key lunar observations accurately even had he wished to. Hooke was Secretary of the RS at that time and Wren and Hoskins were to be the next two presidents.

As Flamsteed was later to bitterly comment *'...soon after Sir Jonas Moore's death, Mr Hook got this* [the mural quadrant] *called for back. It was returned and lies useless ever since in the repository of the Royal Society.'* [16] Whose name did the RS intend to promote had Flamsteed resigned? Certainly not Wren who could have had the appointment for the asking from the outset and certainly not Hooke, who would never have been interested in such a long-term commitment so far from his nocturnal haunts. Who else but the young gifted and enthusiastic Edmond Halley could have fitted the bill?

If so, that plot also failed and Flamsteed was to remain in his post for another 40 years; all the while making observations of his choice as best he could. He tried to replace the mural quadrant with an ambitious 140° mural arc, which would have enabled him to map every visible star. It was made by Tower of London workers and paid for from his own pocket, but it had to be scrapped because it was too flimsy.

If Flamsteed had not been so intent on completing God's mission and had settled for a more modest quadrant, all might have boded well for the observatory and the longitude quest. As it was, for the first 13 years of its existence the observatory was without a decent fixed mural angle-measuring device of any kind [17].

His miserable salary was another reason for Flamsteed's failure to carry out his remit. Not only had the king ordered him to teach mathematics and navigation to two pupils from Christ's Hospital School in his spare time [18], but he had to take on other fee-paying pupils to make ends meet. In 1684 he became rector of the rural parish of Burstow over 20 miles to the south which brought in a much-needed additional £40 (£8,000) a year.

This sealed the fate of any lunar data project because each year at the time of the autumn equinox he had to spend two weeks in Burstow supervising the annual collection of tithes and often as not also spent at least one week away from the observatory at Christmas, the time of the winter solstice. Although he employed assistants at the observatory, they were not able or willing to take these essential lunar measurements in his absence. So because of his king's miserliness and the removal of the mural quadrant by the RS, Flamsteed was free to concentrate on his private project using his big sextant.

Soon after this aborted attempt to oust the Astronomer Royal, 'some persons' first began circulating a specific rumour that continued to dog Halley for the rest of his life (see also chapter 25). Halley had, so they claimed, cuckolded Hevelius *'by lying with his Wife when he was in Dantzig, the said Hevelius having a very pretty Woman for his Wife, who had a great kindness for Mr Halley and was observed often to be familiar with him.'* [19] Flamsteed was apparently one of those 'persons'.

On a spring day in 1682, the 25-year-old Edmond Halley married Mary Tooke, the daughter of an eminent London lawyer in what was apparently a marriage of convenience. Members of her wealthy family were shareholders in shipping companies which, like all marine ventures of the era were involved directly or indirectly in the slave trade.

The young couple set up house in the fashionable north London suburb of Islington, above and away from the noise and bustle of the city. It was here in the cleaner environment that Halley installed his astronomical instruments. From the Royal Greenwich Observatory's historical records, it seems that for the first three years of his married life he spent most nights looking through his telescope! Ironically almost nothing is known of the couple's 54 years of life together other than two minor comments from Flamsteed's diaries.

Significantly, most of Halley's observations were of the Moon, many of the measurements relating to appulses and occultations. One classic early example was the occultation of Porrima, a 3.45 magnitude star in the constellation Virgo by a fine crescent Moon on 28th October 1682 *. According to Halley's first (anonymous) biographer, he then began a regular course of lunar observations stretching from November 1682 to March 1684. Originally it had been his intention *'to carry on these observations diligently...through the whole Course of 223 Lunations ... but he was interrupted at the time last mention'd, by unforseen domestic occasions, which obliged him then to postpone all other considerations, to that of the defence of his Patrimony.'* [20] Halley's father's body had just been washed up in the tidal reaches of the Medway river. He had apparently been murdered and having died intestate, Halley's stepmother was trying to claim his rightful inheritance.

By the time of his father's murder Halley must already have amassed more information on the Moon and its passing encounters with stars than anyone (over 200 detailed observations) in the entire history of astronomy to date, but he had also begun to carefully note angles and distances between the prominent stars that made up local groups adjacent to the ecliptic. This would one day allow him to estimate longitude even when the Moon drifted 'backwards' through a group of stars 'missing' all of them. It was also from this private observatory that he made the observations of the 1682 Great Comet, which helped him to predict its return in 1757 or 1758. When it did return 15 years after his death, it was immediately designated 'Halley's' comet; the 1066 comet depicted in the Bayeux tapestry and possibly first noted in 1059BC.

(1) Eugene MacPike, *Correspondence & Papers of Edmond Halley* (Oxford, 1932), p.37.

(2) Francis Baily, *An Account of the Revd. John Flamsteed* (London, 1835/1966), p.xxv.
(3) Colin Ronan, *Edmond Halley; Genius in Eclipse* (London, 1970), p.35.
(4) F. Baily. p.667.
(5) M. Feingold, *John Flamsteed and the Royal Society/Flamsteed's Stars* (London, 1997), p.38.
(6) F. Baily. p.668.
(7) E. MacPike. p.40.
(8) Thomas Birch,*The History of the Royal Society of London* (London, 1756/7),Vol. 3, p.433.
(9) E. Forbes, L. Murdin & F. Willmoth, eds., *The Correspondence of John Flamsteed* (Bristol, 1995),Vol. 1, p.642.
(10) F. Baily. p.116.
(11) T. Birch. Vol. 3, pp.441-2.
(12) F. Baily. p.lix.
(13) E. Forbes, L. Murdin & F. Willmoth, eds. Vol. 1, pp.652-3.
(14) E. MacPike. p.43. (15) Ibid., p.42.
(16) F. Baily. p.45.
(17) Allan Chapman, *Dividing the Circle* (London, 1990), p.54.
(18) F. Baily. p.xxvii.
(19) Alan Cook, *Edmond Halley* (Oxford, 1998), p.326.
(20) E. MacPike. p.6.

CHAPTER 8

Hundreds of books and many millions of words have been written on the life of Isaac Newton and a poll taken some years ago ranked him second only to Muhammad as being the most influential person in history, one place ahead of Jesus Christ! The results would have horrified him because he seems to have placed himself firmly one place behind Jesus early in his career, a conviction that had a direct effect on every other major character in this story.

Sometime during October 1642, a short six months after his marriage, Isaac Newton senior died aged 36, leaving his heavily pregnant wife Hannah to cope with the family farm as best she could.

On Christmas Day (according to the old Julian calendar then in force) her son was born, a sickly baby thought unlikely to survive and so minute he could, so rumour has it, have fitted into a quart pot. Isaac junior somehow survived his first harsh Lincolnshire winter but when he was three years old, his mother re-married a rich vicar more than twice her age. Her toddler son was not part of the bargain and was abandoned, remaining at the family home in the care of his illiterate grandmother. His mother's occasional visits with one or other of Isaac's new half-siblings in tow were random and merely served to further disturb an already confused child.

When Isaac was 11 and still small for his age, his stepfather died, his mother returned to the family home together with two young daughters and another son and immediately sent her firstborn away to school where he was lodged with an apothecary. At school he was given no formal mathematical training, was bullied as so often happens to misfits, and learned more from the kindly pharmacist than he initially did from his teachers. Eventually his headmaster glimpsed the latent talent lurking somewhere deep within this introverted adolescent and groomed him for university, but on the point of being recommended for Cambridge his mother selfishly removed him from school so that he could help run the family farm.

Isaac junior's foray into farming was a deliberately contrived short-lived disaster; he had no intention of becoming a nobody farmer and would rather take his chances with the school bullies. A frustrated Hannah allowed him to return to school, but only after being bribed to do so by the headmaster, and within a year Isaac was accepted for Trinity College Cambridge; not that she cared.

This 19-year-old Cambridge University freshman's notebook contains interesting insights into that disturbed childhood. In it Newton had drawn up a list of the sins he had accumulated which included *'threatening my father and mother ... to burne them and the house over them'* and *'punching my sister'.* [1] At Cambridge he became a loan shark in order to make ends meet, displayed a remarkable talent for grasping scientific fundamentals and soon understood considerably more about mathematics than his tutors. He took in a lodger John Wickens to help defray expenses; an extremely close relationship that lasted 20 years and set a few tongues wagging.

Newton graduated with a second-class Bachelor of Arts degree in 1665, which was just sufficient to enable the 22-year-old to remain at Trinity College. There was a break in his studies when the plague reached Cambridge and it was at this time that Newton began to give serious thought to the monumental problem of exactly what held heavenly objects in orbit. Back at Cambridge he converted his degree into an Master of Arts without difficulty by paying brief attention to the syllabus; a fellowship followed and Newton was free to lay the groundwork for his scientific ambitions, not that he was sure exactly what these were at this stage.

He had been assailed continually by religious guilt since becoming an undergraduate and had begun searching biblical texts for guidance because of concerns over his homosexual leanings, which he believed, were sinful. Although he failed to unearth anything to give him comfort on that count, he did discover a flaw in the basic Catholic belief in the Holy Trinity; that God was the Father, the Son and the Holy Ghost all rolled into one and a belief adopted by the all-powerful Anglican Church.

Trinitarianism, so Newton discovered, had apparently been introduced three centuries after Christ's crucifixion by the Bishop of Alexandria in order to discredit a local priest by the name of Arius who maintained that according to original accounts, God and Christ were separate beings. Newton concluded that Arius was right and took exception to the unethical manner adopted by the bishop's followers to discredit the priest and enforce a belief in the Holy Trinity.

Trinitarianism was clearly a falsehood and Newton could now happily accept his subconscious conviction that he was quite possibly, like Jesus, another of God's earthly creations. In Newton's case, his mission was to explain his Master's universe to the teeming masses in a logical manner. Now, at last he had a clear objective that would perhaps permit him to live with his disquieting sexuality; possibly he might yet discover that the bible was flawed on that account also? But in England in the late 1660's religious tolerance was non-existent and flouting one's Arian (Unitarian) beliefs in public was not to be recommended.

In addition to his detailed biblical studies, Newton delved into the puzzling world of alchemy, the occult and other proscribed subjects. It was this insatiable thirst for knowledge that led him into a trap. He began taking the stagecoach from Cambridge to London, secretly visiting a bookshop at *The Sign of the Pelican* in Little Britain, a stone's throw from the burnt-out shell of St. Paul's cathedral. This bookshop, run by William Cooper, specialised in selling and lending 'under the counter' publications, books which Newton either purchased or copied. It was at *The Sign of the Pelican* that he met many other men with dangerously unconventional opinions, including several who held influential positions and all of whom used aliases for safety's sake when communicating with one another.

Newton's notebooks from this period use initials and never reveal any identities apart from his own pseudonym, which was *'One Holy God'* ; thus betraying his Arian principles in denying the Holy Trinity. These uncomfortable coach journeys to and from London seem to have occurred in 1668 or 1669 and to have ceased abruptly at about the time of his appointment to the Lucasian chair of Mathematics.

Newton was now obliged to give an undertaking to Trinity College to take holy orders. He had already given similar promises to publicly embrace Trinitarianism, the 'true religion of Christ' on two previous occasions during his career at Trinity without apparently giving the undertaking much thought, but now he had been forced into a corner. In return for the prestigious position - an essential stepping stone if he was to develop the complex mathematics required to explain God's universe - he had made a solemn promise to take those holy orders. However, in all conscience, as an Arian he could not possibly seek ordination. He would have to become so indispensable to Trinity College that he could obtain dispensation *and* retain the Lucasian chair.

Divine guidance, in which like most people of his era he was a firm believer, now came to his aid. Divine guidance would soon also aid Flamsteed, the orthodox vicar as he sat in on that Royal Commission investigating St. Pierre.

So Newton set about the urgent task of avoiding his undertaking, not that anyone other than his friend Wickens could have realised what he was doing. He used his metallurgical skills to manufacture the world's first working reflector telescope and worked night and day to develop his theories on light and colours.

Isaac Barrow the first Lucasian professor of mathematics at Trinity before becoming the King's Chaplain, had kept abreast of Newton's work and the latter required little persuading to let him bring the little telescope to the attention of both the Royal Society and the king. This tiny instrument (figure 3), capable of a 38-x magnification, was delivered to the RS in December 1671; by all accounts it was a beautiful example of the instrument-maker's craft. Yet it had been built from scratch by Newton, working with specialist tools that he had made himself. Casting, grinding and polishing the little concave metal reflector mirror required skills few artisans acquired in a lifetime, yet against all the odds an unknown Cambridge mathematics professor had succeeded in a task that had for years defeated Europe's finest craftsmen. It is not known if Wickens made any significant contribution.

What was so special about a six-inch reflector telescope? The often badly distorted images in the normal refracting telescopes of the day were caused because imperfect lenses split the incoming white light into its different colours (chromatic aberration). A telescope without glass lenses would solve this problem.

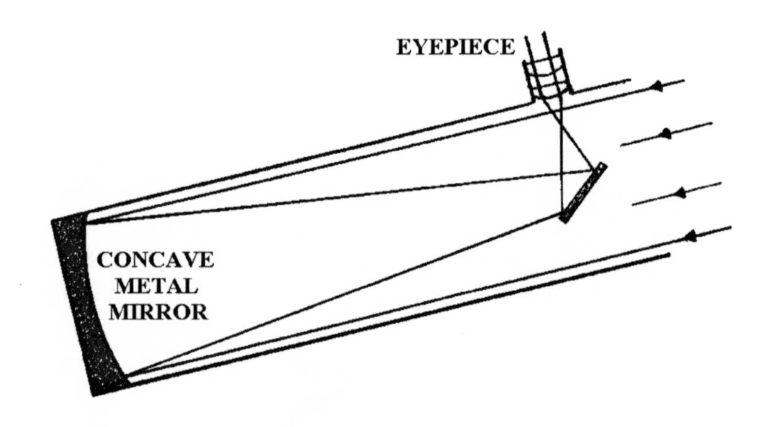

Figure 3. Newton's reflector telescope.
Light rays enter the tube, are reflected back up off the concave mirror onto the small central angled mirror and into the eyepiece.

The society's patron, Charles II was given a personal demonstration and although no one thought to invite Newton to the ceremony, he was invited to become a fellow of the RS and even more importantly, his name had become known in court circles. Typically Newton did not pursue the commercial possibilities; the instrument had served its purpose and his tiny masterpiece soon became ineffective because the wonderful concave metal mirror was allowed to tarnish.

Sadly Hooke had made some unfortunate comments at the RS meeting at which Newton's telescope had been inspected. He had claimed that not only had he manufactured a similar instrument back in 1664 but it was only *'about an inch long ... which performs more than any telescope of 50 foot long made after the common manner.; but the Plague happening, which caused his absence, and the fire, which demanded his employments about the City, he neglected to persecute the same, being unwilling the glass grinders should know anything of secret.'* [2] No one took him seriously but such childish behaviour was totally unwarranted and Hooke failed to produce the instrument.

Others also voiced criticism, mistaking Newton's motives and claiming prior invention. However, the Lucasian professor had never implied the reflection principle was his idea; he had simply managed to manufacture a working model. Unaware of Hooke's sarcastic 'put down' Newton then felt confident enough to declare the true purpose of his foray into the scientific instrument manufacturing field by writing a long letter to Henry Oldenburg, Secretary to the RS outlining his experiments on light and colours.

Unfortunately, Newton's hypothesis was flawed in parts and was at odds with Hooke's own ideas that were also flawed in parts. Within days his letter had been read at a society meeting and within a fortnight Hooke had pronounced judgement that was unfair but not offensive to a disinterested party, although from Newton's standpoint it certainly was. The eventual outcome of this famous feud and Newton's belated discovery of Hooke's childish response to his reflector telescope, resulted in the world being deprived of Newton's numerous and groundbreaking optical experimental results for three decades. He would submit no further papers on the subject to the RS - finito. In any case, he was now pre-occupied with the next stage of his mission.

In 1674, matters finally came to a head on the religious front when Newton found himself being pressed to affirm his orthodoxy and take holy orders. However having been wise enough to keep his Arian beliefs to himself, he now called on his mentor Barrow (a strict Trinitarian) for help. Barrow not wishing his old college to lose such a brilliant scientist, paved the way for Newton to apply directly to Charles II for dispensation and by April 1675 the king had granted Newton and all subsequent holders of the Lucasian professorship exemption from taking those holy orders.

Hooke's insistence on constant argumentative coffee house style pronouncements did a second disservice to science in that Newton's fragile mental state now suffered a serious setback. Pressured also by other scientific discussions/disputes with the mathematician Gottfried Leibnitz and Anthony Lucas the English Jesuit and critic of his optical claims, Newton became more and more irrational. Soon he was convinced that his opponents, encouraged by papists, were trying to undermine his research by engaging him in futile correspondence. His correspondence with Leibniz in the summer of 1676 had included unbreakable enciphered passages that apparently hid essential sections of the development of his new mathematical methods (chapter 26).

Leibniz had already developed a similar mathematical tool but could not discuss this because of Newton's secretiveness. Newton unwisely may have included at least one disguised reference to his Arian beliefs. One of the anagrams (6a cc d æ13e ff 7i 3 1 9n 4o 4q rr 4s 8t 12v x) included a central section i3l9n. AD319 was the year that Arius travelled to visit Eusebius after being accused of heresy and which gave rise to the Arian controversy.

Then in the winter of 1677 the little laboratory he had constructed when his experiments had began intruding into his living space burned down, destroying most of his research papers. It was too much for Newton to handle and for several months he was verging on the certifiable. Although Wickens did his best to help him through this period of intense despondency and frustration, their relationship began to cool.

Suddenly Newton snapped out of his depression. Having unearthed the flaws in Trinitarianism through carefully studying biblical texts, Newton now became convinced, like many others, that the Bible also held the keys to the hitherto uncertain dates of historical events.

A brilliant mathematician possessing special God-given gifts could surely convert some of the events *forecast* in the Bible into actual dates. And if a few happy coincidences were to confirm that gift, why anything was possible surely? Having been born with no father prematurely on Christ's birthday, and then to survive despite being so minute that he could be fitted into that quart pot, must have reinforced Newton's faith in his abilities; even if he had been the one to mention the pot and the premature claim which avoided the embarrassing likelihood of being conceived out of wedlock. Again remarkably similar conclusions to Flamsteed's belief that God had permitted him to survive for a special purpose despite his afflictions.

Although much of Newton's biblical research was commenced in the 1670's he continued with these studies up to the very month of his death in 1727, yet published nothing. Seemingly a blind alley although Newton burned boxes of documents weeks before his death and hundreds of scribbled notes relating to his biblical research still wait to be properly scrutinised three centuries later. At one stage Newton predicted that the great tribulation of the Jews would end in 1944, to be followed four years later by the second coming of Christ. Armageddon would occur sometime during the 21st century but he seems to have put that date back by about 1,300 years at one point.

To many of us such prophesies are merely fortune telling, lacking scientific merit, however convincing, but exploring the boundaries of science all the while risking being branded a heretic is a far fairer description of Newton's biblical forays. This facet of Newton's life is only highlighted because his deep and long-lasting interest in numerology takes centre-stage later in this story.

However, the second of Newton's progressive ideas ran into a major obstacle of a different nature. Many people believe that Newton suddenly discovered 'gravity' whilst sitting under an apple tree and puzzling over why an apple should fall to the ground yet the Moon remain hanging in space.

In reality four of London's residents, Hooke, Wren, Halley and Flamsteed (when not otherwise pre-occupied) were all attempting to define the laws which governed the motions of heavenly bodies; to explain the parameters of the strength of the glue which held the planets and attendant moons in their orbits round the Sun and prevented them from flying off into space, crashing into one another, coming to a stop or being sucked onto the super-magnet, the Sun. And they were each in his different way making progress, but progress from the viewpoint of astronomers first and mathematicians second. Newton was now working to define these orbits mathematically, not having the time or inclination to peer through a telescope half the night and probably seeing little point in doing so as long as he could persuade others to do this for him.

In 1679, Newton's mother Hannah died, Newton inherited the family estate and he took time out to sort out his finances; a similar situation that Halley was to find himself in following his father's death five years later.

Back at Cambridge Newton again clashed with Hooke who involved him in correspondence regarding views on the behaviour of falling bodies, correspondence that Newton considered to be private discussions between scientists.

However, Hooke gleefully publicised a flaw in Newton's mathematics and then proclaimed that he himself had just solved the problem of planetary motion but refused to tell anyone how he had done so. Once again Newton retreated into his shell.

In the summer of 1684, three months after his father's funeral the innately curious Halley, having had enough of the boastful Hooke and the small-minded Flamsteed, paid the reclusive Lucasian professor a visit.

Who knows what the 28-year-old thought of the heavy dark red drapes and furnishings of Newton's claustrophobic rooms or of the man's odd mannerisms; Halley was not one to gossip. He bravely asked Newton point-blank if he could provide a mathematical explanation that in truth had for years proved beyond the abilities of everyone else. He was somewhat surprised to be given a 'Hooke-type' reply; yes Newton could, but he could not lay his hands on the papers at present. When he had located them he would deliver a mathematical demonstration on the inverse square law of planetary motion; the answer to the puzzle. Halley returned home a little disappointed but diplomatically said nothing of his encounter or of Newton's stunning claim.

During Halley's patient wait, caused not by Newton being unable to locate the papers but because, after his unfortunate clashes with Hooke he was preparing his case meticulously, Newton decided to write to Flamsteed seeking certain astronomical data to assist his research.

Flamsteed whom he had met briefly at Cambridge 10 years earlier whilst the latter was obtaining his degree, was helpful, supplying most of the data but was unable to comprehend the underlying reasons for the requests, Newton not having bothered to offer proper explanations.

In fact, Newton had already computed the orbit of Saturn and come to the conclusion that what little data was available did not fit his figures. Therefore that data must be erroneous! For example instead of explaining his reasons to Flamsteed for requiring orbital data of Saturn he simply wrote *'This planet so oft as he is in conjunction with Jupiter ought (by reason of Jupiter's action upon him) to run beyond his orbit about one or two of the Sun's semi-diameters or a little more and almost all the rest of his motion to run as much or more within it'.* [3] In other words the gravitational attraction of one little speck in the night sky could alter the orbit of another little spec; a mind-boggling revolutionary statement which was as perfectly obvious to Newton as it was incomprehensible to Flamsteed.

Halley's response to such a suggestion would have been to rush off and run some practical checks. Flamsteed merely replied *'It seems unlikely such small bodies as they are compared to the Sun, the largest and most vigorous magnet of our system, should have any influence upon each other at so great a distance'* [4] which merely highlighted his lack of vision.

The correspondence ceased abruptly; Newton could not be bothered with providing explanations to one he considered so hidebound. Flamsteed could read about it in due course, but the fact that Newton was to be proved right once Flamsteed did get round to checking for himself, would do nothing to improve relations between them, or for that matter with Hooke who was even now still trying to make good his rash boast.

Meanwhile Halley heard nothing, but knowing a little of Newton's psychological make up, waited patiently. Suddenly out of the blue arrived a hand-delivered nine-page document titled *'De Motu Corporum in Gyrum - On the Motion of Revolving Bodies.'* Realising the paper's significance Halley returned to Cambridge to obtain permission to read the document at the next RS meeting; another point in his favour. Hooke would have simply revealed the details and either tried to belittle the author or lodged a prior claim. Newton agreed and then the RS asked permission to publish but Newton baulked at the suggestion, already now long on the way (as evidence his comments to Flamsteed) to using the *'De Motu'* material as a base for a unified theory which explained everything rather than merely the motion of revolving bodies.

A year later in February 1686 Halley was elected Clerk to the Royal Society, a supposedly paid employment that necessitated his resignation as a fellow and saved him his annual membership fee. Hooke had been one of the unsuccessful candidates. If Halley had expected his financial position to immediately improve he was in for a shock; his salary was not to be finalised until he had been in office for a year, when it was settled at £50 per annum plus a gratuity of a further £20. In modern terms a modest £14,000 for doing a very demanding job. However as the society was strapped for cash despite having sold Chelsea College for £1,300 (£260,000) a few years earlier, Halley was paid in kind; 70 copies of *'Historia piscium - The history of Fishes'* which the society had failed to sell.

Halley then persuaded Newton to publish all his material and the RS eventually voted in favour of sponsorship. This despite a well-orchestrated protest from Hooke and his supporters, which was neatly overcome by giving their new Clerk the financial responsibility. Halley was now obliged to find the money to pay the printers. On the other hand, any profits would be his, but it is doubtful if there were any, as he had to bear the costs of 100 expensively bound presentation copies.

But few could comprehend the publication because it had deliberately been laid out by Newton in such a way that the reader had to be able to fully understand one highly technical proposition before there was any point in reading the next; and all written in classical Latin.

Despite this and the furore created by Hooke because any references to his contribution had been deleted at the last minute by a vindictive Newton still seething at his protest lobby, the publication firmly established Newton's reputation. He never seems to have considered Halley's contribution of importance and whilst there is no evidence that the skinflint was invited to contribute to the printing costs this is in all probability because Halley dare not ask; getting the temperamental recluse to agree to publication had been difficult enough.

Philosophiae naturalis principia mathematica (Principia) contained the underlying mathematical proofs, which could explain, and be used to predict, the continually varying speeds and shapes of the elliptical orbits of all the bodies orbiting the Sun, together with the manner in which they would interact. Nevertheless, Newton was painfully aware that he still could not predict the lunar orbit. Until he could do so, his unified theory was incomplete and his proposition on the theory of lunar motion as set out in *'Principia'* was indeed nothing more than a proposition.

(1) Michael White, *Isaac Newton; The Last Sorcerer* (London, 1997), p17.
(2) H. Turnbull, ed., *The Correspondence of Isaac Newton* (Cambridge, 1959-60), Vol. 1, p.4.
(3) Ibid., Vol. 2, p.407. (4) Ibid., Vol. 2, p.409.

CHAPTER 9

Galileo's discovery of four moons orbiting Jupiter had opened the way for determining longitude on dry land. Initially it had seemed all too easy, and in one way it was. Io, then considered Jupiter's closest satellite, buzzed round the planet once every 42½ hours. Each orbit contained four separate specific observable events - when Io suddenly disappeared behind Jupiter, when the moon re-appeared again on the far side equally suddenly, when it started to pass in front of Jupiter and disappeared from sight when it began to cross the permanently sunlit face of the planet and when it cleared the face again.

Most nights that Jupiter was visible a time-able event was occurring. No matter exactly when these events were taking place, each was occurring at almost exactly the same time wherever the location of the observer.

So, as the French had been very quick to realise, if there was a central observatory continually registering the exact time of every possible event, observers elsewhere merely had to take the local time of an 'Io' event and check this against the central data bank in order to determine the precise longitude of their observation post. No matter that it might well take a distant local observer several months to obtain the results. At this stage, not a lot of help to a mariner lost in a vast ocean, but at least he may now know where the coastline was supposed to be on his chart.

In 1679, Jean Picard and Phillipe de La Hire, using this method had managed to map the true outlines of France. On seeing the results, which reduced the size of France considerably, Louis XIV famously pointed out that he had lost more territory to his astronomers than to his enemies[1].

However, not all Jovian events were observable from Paris or any other single collection point. An event timed in Java could not be recorded in Paris half a world away because Jupiter was below the horizon in Paris at the time. Prediction tables were required but the difference between random timing of observable events and advance prediction required a giant advance in astronomical knowledge.

The biggest stumbling block was caused because both Jupiter and our observation platform (Earth) are continually on the move and both planets have different shaped elliptical orbits round the Sun, which are not in synchronisation. This means that the angle that Jupiter and its attendant moons present to Earth is continually altering. Without a complete computation of the relative positions and angles of the two planets and several other points (the speed of light for example), advance prediction tables could not be produced.

Once Flamsteed realised that Halley was serious about the prospects of determining longitude on the high seas by the lunar occultation/appulse method he must have thought long and hard about how to avoid losing face if Halley were to be successful. Working *with* Halley never seemed to have occurred to him and he certainly could not compete openly owing to the dearth of lunar data in his possession and the lack of prospects of ever obtaining enough without jeopardising his private project. So Flamsteed decided to actively promote an alternative method of determining longitude, which would not take up too much of his time.

Cassini, by all accounts was nowhere near to producing a Jovian almanac. Flamsteed did not even need a mural arc; Jupiter's orbital data could be collected and then predicted using his big sextant and the timing of the moons could be accomplished with his long telescope and pendulum clocks. He was well aware of the practical snags, the biggest being how anyone could make accurate observations from the rolling deck of a ship, but being no sailor this problem was perhaps not readily obvious. In any case that snag appeared to a landlubber to be similar to the one Halley would face with his lunar occultation/appulse method.

From about the time that Halley had commenced regular lunar observations from his house in north London in 1683 and Flamsteed realised he really had a rival, the Astronomer Royal had begun promoting the Jovian moons alternative. In all he was to publish a raft of papers on this topic (including two in Latin for the benefit of foreigners) in *Philosophical Transaction* between the years 1683 and 1686 [2].

Considering Flamsteed's reluctance to publish his stellar and lunar observations, his behaviour in regard to hastily publishing Jovian information and almanacs tends to confirm he was aware of the need to work fast to beat Halley. His project to map the heavens on behalf of his Maker fell into an entirely different category; naturally, in that case nothing less than perfection was acceptable.

Indeed he was in such a hurry to publish his Jovian almanac for 1684 that he never even corrected the proofs, thus confusing the issue somewhat; *'A letter from Mr. Flamsteed concerning the Eclipses of Saturn's Satellit's for the year following. 1684 with a Catalogue of them, and informations concerning its use.'* Typically, the Astronomer Royal used the opportunity to take a dig at Halley in particular and seamen in general.

What mariners made of the title's typographical mistake - Saturn instead of Jupiter - when they were hard put to locate either, is anyone's guess. What they thought of the following is sadly all too obvious. *'And I must confess it is some part of my design, to make our more knowing Seamen ashamed of that refuge of Ignorance, their Idle and Impudent assertion* that the longitude is not to be found, *by offering them an expedient that will assuredly afford it, if their Ignorance, Sloth, Covetousness, or Ill-nature, forbid them not to make use of what is proposed.'* [3]

In dealing with Halley, and incidentally undermining his own enthusiastic comments to the commission which had resulted in the establishment of the Royal Observatory, Flamsteed had this to say: *'Those of them that pretend to a greater talent of Skill than others, will acknowledge that it might be attained by Observations of the* Moon, *if we had* Tables *that would answer her* Motions *exactly; but after 2000 years experience (for we have some Observations of* Eclipses *much ancienter) we find the best* Tables *extant erring sometimes 12 Minutes* [of arc] *or more in her apparent place, which would cause a fault of half an hour, or 7½ degrees in the longitude deduced by comparing her place in the heavens with that given in the* Tables *: I undervalue not this Method, for I have made it my business, and have succeeded in it, to get a large stock of good* Lunar observations *in order to the correction of her* Theory, *as the ground work for better* Tables *but the examination will be the work of a long time.....'* [3]

In other words, the Jupiter's moons method was the one to opt for after all. As for the *'large stock of good lunar observations'* claim, Flamsteed, at the time of writing did not have one single positional lunar observation he was absolutely sure of.

The instructions for using that 1684 Jovian almanac were muddled but most of the data were reasonably accurate. Flamsteed acknowledged Cassini's help and also had the good grace to mention that *'if it not be practicable at Sea they cannot deny but it is at land; That the true* longitude *of remote* Coasts *from us are the first thing desired for the correction of their* Charts...*'*

Sensibly, there was no mention of the fact that Halley had spent a year on St. Helena with top of the range equipment without managing the feat once. He continued *'let them attempt these first, and I doubt not but the success will encourage them so much, that they will readily find means to put it into practice at Sea.'* [4] Wishful thinking and in any case there were one or two dangerous errors/mistakes in his 1684 almanac.

Flamsteed published three more annuals covering the years 1685-7. His 1687 version contained some more obvious ghastly typographical errors, which the newly appointed Clerk to the society (Halley) must surely have noticed and should have corrected. This publication also had an unfortunate misprint in the title, suggesting the article contained the results of a lunar total eclipse observed in Lisbon nine years hence in 1695. This error was properly corrected to 1685 in one place in the text but 1686 was mentioned elsewhere. Flamsteed also admitted that his last published set of tables contained errors in relation to the timing of the 2nd and 3rd moons but hoped that in two or three years time such errors would be corrected [5].

The Astronomer Royal had rightly warned in 1685 that the data contained in his 1686 almanac were still suspect; he had not yet fully mastered the mathematical intricacies[6]. Newton's comments on the gravitational effects of Saturn on Jupiter's orbit, which Flamsteed could not grasp, coming as they did at this point, might possibly have prompted this admission.

All in all 5 out of 10 for effort but seamen who take 50-50 chances die young. And if Halley had turned a blind eye to the typographical errors, who can blame him?

(1) Albert Van Helden in *The Quest for Longitude* (Harvard, 1996), p.95.
(2) John Flamsteed, (R.S. Phil. Trans., London, 1683-1686), Vol. 13, pp.322-3; 401-415; Vol. 14, pp760- 765; Vol. 15, pp.1215-1225; Vol. 16, pp.196-206.
(3) Ibid., Vol. 13, p.405 .(4) Ibid., Vol. 13, p.407. (5) Ibid., Vol. 16, pp.196-206. (6) Ibid., Vol. 15, pp. 1216.

CHAPTER 10

John Shovell and Anne Jenkinson set up house in the Norfolk hamlet of Cockthorpe a few miles from the village where Horatio Nelson would be born a little over 100 years later. There they produced a batch of offspring from about 1645 onwards. The son they named Cloudesley (the surname of his maternal great-grandfather) was born sometime in early December 1650. Cloudesley Shovell; two names that, over the years would be spelt in nearly as many different ways as William Shakespeare's surname and whose body would one day be laid to rest close to the bard's memorial in Westminster Abbey. If this story must have a villain to balance the heroic qualities of Halley, Cloudesley Shovell is the man who best fits the part.

Cloudesley was the only one of John and Anne's children to survive into adulthood possibly because he was fortunate enough to join the Royal Navy at the age of about nine under the sponsorship of a family friend Vice Admiral Sir Christopher Myngs. The boy sailor thus avoided the various diseases that killed off a goodly percentage of the resident English population in the mid 17th century. A lucky start and similar good fortune was to attend him, if not his close associates, throughout almost his entire colourful naval career. Almost but not quite.

Shovell's first fleet action had been the Battle of Lowestoft in 1665 but at the 'Four days fight' off North Foreland in June 1666 Myngs, directing his sector of the line from the deck of *Victory,* a second-rate battleship carrying 82 guns, was shot in the throat and neck by sharpshooters and died two days later. Not particularly noteworthy amongst all the carnage but young Shovell and *Victory's* first lieutenant John Narbrough who would take over responsibility for the boy's further naval education, both witnessed the incident, and with Myngs' death both lost their patron.

There certainly must have been something special in the soil or on the salt air of that particular coastal strip of north Norfolk where Shovell was born. Narbrough was from the same hamlet, Myngs was from Salthouse only five miles distant, and Narbrough and Shovell were to follow in Myngs' footsteps and also become knighted Admirals of the English Royal Navy.

The 'St James day fight' in August found young Shovell under the protective wing of Narbrough aboard the 60-gun *Assurance.* It was from the deck of this warship that he witnessed the torching of Terschelling. The following year a powerful English squadron, which included *Assurance,* sailed for the West Indies with instructions to attack the French Caribbean fleet, forts or anything else they could savage. *Assurance* was still commanded by Narbrough who again took Shovell with him.

Sailing south to the mainland of South America searching for bounty in the French settlement of Cayenne they landed troops, chased off the Governor, broke into a liquor store, became rolling drunk and fired the town. 36 guns from the fort, 150 Negro slaves and quantities of copper, sugar and tobacco were among the booty.

The captured guns and the bulk of any prize treasure was by law the property of the state (if the special commissioners could lay their hands on it) with the remainder split between the fleet, with the Admiral and the General in charge of the Marines taking half between them. 16-year-old Shovell received a Negro in lieu of prize money, a poor wretch who was probably later sold in the West Indies for about £15 (£3,000).

The sacking of Cayenne was the start of Shovell's rise to fame and the small sum received for that slave gave no hint of what was to follow. During the course of his naval career, Shovell would somehow accumulate nigh on £1 million (£200 million).

During 1672 Narbrough was, for a time the first lieutenant aboard the new first-rate 100-gun battleship *Prince,* with Shovell, now a 21-year-old midshipman still in close attendance. This drop in rank for Narbrough was in reality a promotion because *Prince* was the chosen flagship of the Duke of York, who was once more allowed to risk his neck at sea. Shovell was about to come in (distant) contact with royalty for the first time when Charles II, his bastard son the Duke of Monmouth and entourage would spend nights aboard the new pride of the fleet. However, that was before the Dutch decided to use *Prince* for target practice.

At the key battle of Sole Bay off the coast of East Anglia, the Dutch singled out the Duke's flagship and did their very best to eliminate the heir to the English throne. During the fight the captain was killed, as were many of the officers and nearly 100 of the crew.

The hapless Matthew Wren was seriously wounded but the Duke of York, Narbrough and Shovell somehow all came through unscathed. Possibly this can be considered to be Shovell's first stroke of good fortune, discounting the slave acquisition.

Another *Prince* survivor was an ordinary seaman by the name of William Dampier, a piratical character who was wounded in the action and would one day meet up with Halley and Shovell. The Duke transferred his pennant to the more seaworthy battleship *St Michael* under the command of the

'Torcher of Terschelling' Sir Robert Holmes, and the Dutch left Narbrough the newly promoted captain to withdraw from the fight as best he could.

The Earl of Sandwich in the *Royal James* had been first to engage the enemy that morning and had also come under intense attack. By lunchtime his ship had been reduced to a blazing wreck and most of the officers and crew blown to bits or drowned, including the unfortunate Earl.

One of the few survivors from that particular floating abattoir was a certain young Captain Richard Haddock who, although wounded escaped by diving through a gun port as a Dutch fire-ship rammed into *Royal James* and exploded. Haddock, accompanied by Shovell would 27 years later come face to face with Halley and then Newton in peculiar circumstances (chapter 18).

The following year, just as the recently formed Royal African Company was getting into its stride, Narbrough was given command of a Royal Naval squadron escorting a large number of merchantmen carrying silks, spices and other valuable produce back to England from Turkey; the 1673 annual Smyrna convoy.

Shovell was by now an assistant to the master navigator and the ship they were sailing in was the 80-gun *Fairfax.* After a long trip through the Mediterranean and up north across the western edge of the Bay of Biscay, they were almost home when they suddenly and unexpectedly found themselves amongst breaking surf in the dead of night. *Fairfax* managed to fire warning shots and hastily drop anchor.

When dawn broke Narbrough found to his horror that half the squadron were in amongst the dangerous Bishop and Clarks' rocks (Bishop Rock) directly to the west of the Scilly Isles. They were extremely fortunate to escape without loss and for Shovell, the navigator's assistant mate, it should have been a chastening experience. Young Shovell's second stroke of luck.

The excuses given for this debacle are worth noting for later; this was not the last time Shovell was to find himself amongst these dangerous rocks.

First, in Narbrough's personal experience, navigators always got further north than their reckoning on coming into the Channel from the south. This was caused by a great in-draught of current that set into the river Severn and the St. George's Channel - of which there was no evidence as this event only occurred during exceptional climatic circumstances, which were certainly not present at the time.

Second, was the constant changing of course whilst shepherding lagging merchantmen who nevertheless all avoided the danger. Finally, the fleet navigators reckoned they were well clear of the Scilly Isles anyway.

Samuel Pepys, following a similar hair-raising experience aboard a Royal Navy vessel, was sufficiently shaken to comment *'It is most plain, from the confusion all these people are in, how to make good their reckonings, even each man with itself, and the nonsensical arguments they would make use of it to do it, and disorder they are in about it, that it is by God's Almighty Providence and great chance, and the wideness of the sea, that there are not a great many more misfortunes and ill chances in navigation than there are.'* [1] (see also chapter 15). Spoken from the heart and when the wideness of the sea was absent, disaster all too often followed. Certainly in the Narbrough incident, as he had admitted he knew of the dangers he surely should have made due allowance.

Most relieved of all were London's merchants because Narbrough's squadron were carrying nearly two million silver pieces of eight naively entrusted to the Royal Navy escort for safekeeping.

During the 1676 Mediterranean campaign, Second Lieutenant Shovell was placed in charge of a small boat attack party by his mentor Narbrough over the heads of many who considered themselves better qualified. Under his inspired leadership and the especial bravery of his friend and commander of another of the boats, James Greeve, they succeeded in burning a number of the Tunisian men-of-war tucked up 'safely' in the inner reaches of the harbour at Tripoli. Although many lost their lives in this action, Shovell escaped with nothing more than a minor wound.

By 1679 Shovell had achieved his own command and in his free time was teaching the Duke of Grafton (an illegitimate son of Charles II by Barbara Villiers who had been displaced as official mistress by 'weeping willow' Louise Keroualle) the arts of seamanship. Shovell's own most spectacular acts of seamanship tended to be reserved for occasions where bounty was available and this ran him into temporary trouble with the Admiralty in 1683 when, by now a squadron commander, he gave permission for one of his ships to return home early from Cadiz. It transpired that this vessel was carrying money the squadron had looted but Shovell managed to convince the Admiralty that *Crown* had been sent home for refit because it was dangerously slow and un-seaworthy.

On the way back to England *Crown* had been requested to heave-to by a Royal Navy frigate to explain why she was sailing alone, which she failed to do; the frigate commander later reporting that *Crown* was very speedy! Again Shovell was in luck; his meteoric career could so easily have been snuffed out over that particular escapade.

There is a body of evidence to suggest that Shovell was a bounty hunter, a slave trader and a moneylender and Narbrough often acted as his London banker. In 1685, Shovell left slaves for sale in the Canary Islands and at one point in the 1680's was owed over £2000 (£400,000) by fellow officers, one of whom was Admiral George Rooke! There could not have been many officers in their 30's that had commenced their life at sea penniless yet were in a position to lend such sums. There were few either who could have found themselves in a position to lend the Duke of Grafton money to buy a watch or a black slave.

Following the death of Charles II, the Duke of York became James II, announced his belief in the Catholic faith and had to flee the country when his own daughter Mary and her Dutch husband and first cousin William deposed him in a more or less bloodless coup.

What undoubtedly helped Shovell's career most at this point was his unqualified allegiance to his new king and queen. Lord High Admiral William III knighted Shovell in 1689 and promoted him to Rear Admiral the following year.

Halley was also about to be employed at sea by his new lord and master, at first sight an odd commission for an astronomer. But persons in high places were beginning to realise Halley was someone special for although he must have disliked the unethical behaviour of some of his scientific colleagues, he was unusually diplomatic in the manner in which he expressed his opinions.

Because he had acquired this trustworthy reputation and despite his inexperience as a mariner, navigator or surveyor, Halley had originally been commissioned to conduct a secret hydrographic survey of the Thames Approaches on behalf of James II. Hardly had he started than the king was deposed. Now William sensibly wanted the entire sea area between his two homes to be properly charted.

Halley completed the survey in his usual quiet and efficient manner and his data were handed to the Admiralty. Some details were even published by the Royal Society in 1689. From Halley's point of view this was simply another job well done and another step up the ladder; he could not hope to determine longitude out in the wild reaches of the world's oceans if he could not command and navigate a ship within the confines of the North Sea. No doubt Halley would have continued with his survey work further south into the English Channel but for the war which broke out with France again. This part of his work was delayed until 1701 and it was only then that his charts were made public (chapter 21).

Admiral Sir Cloudesley Shovell's first duty in the new year of 1691 was to escort William III to Holland. The king embarked in his yacht (an armed frigate) at Gravesend and the fleet set about sailing the 150 miles across the southern North Sea to Holland. What with being becalmed and fogbound, it was five days before they arrived in the shallow waters of the Schelde estuary; or so they thought. As usual, the consensus resulted in a confused, disunited conclusion.

The very shallow water at least confirmed they were certainly close to land even if they could not sight it through the swirling fog. Impatient to be home, William and his entourage set off in a rowing boat with a small crew of sailors in what they hoped was the right direction. Eventually someone was persuaded to volunteer to swim for it and to light a signal fire if he survived. He did as ordered but the king wisely preferred to remain in the boat and did not set foot on dry land until the next morning, when he discovered he was on the island of Goree 30 miles from the intended landfall.

Exactly what Shovell and the rest of the escort were doing while all this was going on was mercifully not recorded. One would have presumed that the Royal Navy would have made use of Halley's potentially life-saving survey work but the debacle over the landing of William at the wrong place suggests this was unlikely.

Shovell instead of receiving a reprimand was actually rewarded with a commission as Major of the First Marine Regiment by a grateful Queen Mary in another of those odd circumstances (chapter 11). Even odder was the decision of William to retain such an incompetent admiral as his regular guardian whenever he wished to travel by sea. Shovell's 4th stroke of luck and had he been a cat, he would now only have five lives remaining.

(1) E. Chapell, ed., *The Tangier Papers of Samuel Pepys* (1935), p.129.

CHAPTER 11

In March 1689, Halley had read a paper to the Royal Society about a method of walking under water [1] that had recently been employed in the West Indies. Air was taken down in a 'diving bell' and the free-swimming divers could take occasional gulps of air from the bell's supply without the need to re-surface. Halley proposed to make an improved version, which would enable men to work from inside the bell that could be moved about underwater on wheels. Air would be replenished by sending it down to the divers in weighted barrels. All in all a somewhat hazardous venture and so far merely theoretical.

Predictably Hooke was moved to comment; *'met...Hally* [who] *read a paper of Walking under Water. the same wth what I shewed ye Society 25 years since'.* [2] They both probably knew Leonardo da Vinci had beaten them all to the idea by a wide margin. Three months after the erroneous landing on the island of Goree by his sovereign, Halley was given the chance to test his diving bell when he was asked to carry out a survey on a Royal African Company wreck in the English Channel. However, this was by no stretch of the imagination your average shipwreck.

The years 1685 (46,066 guineas) and 1688 (39,371 guineas) had been the two most profitable years of gold imports into England in the RAC's entire history. In the 1686-7 season 10,815 slaves were sold in the West Indies for £165,000 (£32 million), 6,223 of them to Jamaica. Never again would the company officially trade so successfully in either gold or slaves. It was at this point that James II had sensibly liquidated his stock in the company before less sensibly flinging the Great Seal of England into the river Thames before fleeing the country.

This year, 1688 had marked the beginning of a lingering and painful death for the company because a reformed English parliament pronounced the RAC guilty of abusing its monopoly from the very outset back in 1673. On re-examining the company's charter, it was discovered that the RAC was obliged to pay compensation to 'interlopers' whose ships and goods they had confiscated, a nicety that had conveniently been overlooked whilst James II had remained in power.

In April 1690 parliament was petitioned by merchants whose ships had been hijacked by the RAC many years earlier and the following month the Governors of the company approved compensation to at least four aggrieved merchants and instructed their auditors to set aside considerable sums against further claims. The company was suddenly to all intents bankrupt and left without working capital. To make matters worse the numbers of slave traders prepared to buy licenses from the company dropped alarmingly; why bother? In that watershed year of 1690 the RAC saw its direct slave trade drop by 90%. The deposed James had divested himself of his shares just in time.

From 1690 onwards when William III became directly involved in running the company, it began concentrating on profitable deals rather than volume sales. The RAC also began reducing its reliance on chartered vessels and began leasing out the least productive sectors of its slave trade monopoly. The idea of owning an entire fleet of ships rather than chartering most of them 'ad hoc' was one of the desperate ideas that the Court of Governors of the RAC promoted in order to stem their alarming drop in trade, bankrupt or no.

Employ first-class captains and crews and pay them well, which should put many of the best slaving captains (of which there were surprisingly few) on their permanent payroll and disadvantage their competitors. But the idea got off to a very dubious and suspicious start at a time when the RS were again about to invest in that company.

In mid-February 1691, the armed frigate *Guynie,* owned by the RAC and on its second trip back to England from West Africa, docked at Falmouth at the western entrance into the English Channel almost certainly by prior arrangement. *Guynie* was carrying a fortune in gold (the property of several Portuguese merchants partly insured by City of London brokers) plus 184 elephant tusks. The captain, William Chantrell sent a dispatch from Falmouth asking for a man-of-war escort on through the English Channel and armed sentries were apparently posted on the dock meanwhile.

The Royal Navy, by accident or design, failed to respond promptly to his request, but when three weeks later Chantrell was finally informed that an escort was being considered, he hastily put to sea and headed up the English Channel towards London unescorted.

The next anyone knew was that Chantrell was in the town of Chichester reporting the loss of his vessel off the Sussex coast near the township of Pagham. There seems to be no record of any bad weather accident, striking of rocks, enemy action or even of any of the crew drowning; nothing. A well-armed seaworthy treasure-laden frigate had simply sunk in water deep enough to preclude a salvage attempt but close enough inshore to allow the entire crew to row ashore.

However, 10 fathoms (about 18 metres) was no longer quite out of reach of divers. Treasurer and founding member of the RS and Deputy Governor of the RAC Abraham Hill and RS fellow and Sub-Governor Sir Gabriel Roberts persuaded Halley to take his new-fangled diving bell contraption, no longer merely an idea, to survey the wreck with a view to salvage. Roberts was the uncle of Robert Nelson, a school friend of Halley's who had accompanied him on a two-season jaunt round Italy and France following Halley's hasty departure from Danzig and the close proximity of Elizabeth Hevelius 10 years earlier. Both knew RS Clerk Halley was entirely trustworthy.

The survey started badly. Chantrell, *Guynie's* captain, advised Halley against diving because of the poor state of the hawsers attached to the bell, hawsers that Halley could see little wrong with. Diving could not safely begin until this and a host of other niggling problems placed in Halley's path had been overcome, by which time a full three months had sped by.

When operations did begin, *Guynie* was found to be lying on her side on a shifting sandy bottom; not the easiest of conditions to deal with. There seems to be no record of Halley's reports although he was later to write that he and his divers had succeeded in entering the wreck. No damage report and no salvage. Other divers, who were, years later employed at the request of the assessors seemingly were no more successful. All that is known is that at one point Halley had his salvage vessel hijacked by pirates hoping to steal any recovered treasure.

During the winter of 1691/2 when presumably salvage operations had been suspended, he read three papers at RS meetings relating to advanced salvage methods. Underwater blasting, the use of heavy lifting tackle and buoyancy casks, diving bell improvements and how better to deliver air to divers [1]. Clearly were his divers to experience difficulty in accessing the holds, Halley was one of the few people possessing the expertise to clear a way into the vessel, but there is no record of his doing so.

A singular 'success' came when a diver was rumoured to have recovered an elephant's 'tooth' from the seabed in the vicinity of the wreck site in March 1695, almost exactly four years to the day after the sinking; by which time Halley was no longer involved. One might presume that the gold and the other 183 elephant's tusks are still somewhere under the sand off the Sussex coast; one might, but no one seems to have invested money in more recent searches.

To the sceptical mind, the gold had been secretly removed when *Guynie* docked in Falmouth. Certainly as Governor and major stockholder in the RAC, the king would have had advance notification of the impending shipment, an insured shipment that did not belong to his company.

Where was Shovell when *Guynie* first conveniently arrived in Falmouth? Almost at the exact time Cantrell requested an escort, he was in London receiving that commission from Queen Mary which placed him in the influential post of Major of the First Marine Regiment; a vacancy that had occurred suddenly because of an inexplicable resignation. However, the appointment, unlike a lieutenant-colonelship, specifically involved the control of regimental *on-shore* duties, as Shovell himself was soon to confirm (chapter 12).

A month after Cantrell's request Shovell was in London marrying Elizabeth the wealthy 32-year-old widow of his mentor and banker Sir John Narbrough who had died in the West Indies in 1688; becoming a caring step-father to her three children, the eldest, 10-year-old Sir John junior. Elizabeth was an heiress to a fortune in her own right and already a very rich woman, and her father, John Hill (no relation to Abraham), who had just been appointed Commissioner of the Navy, was a stockholder in the RAC. More than time enough between the two events to arrange for the secret removal of the gold from *Guynie* in Falmouth by a small detachment of Shovell's new regiment of marines on behalf of William III who was in desperate need of extra funds.

If the treasure had been removed either under Shovell's direct supervision or by other means before the sinking, those involved must have been concerned when Halley anchored above the wreck and lowered his diving bell to the seabed for the first time. This would explain the difficulties Cantrell put in Halley's way, the absence of a salvage report and might also account for Shovell's later antipathy towards Halley. If Halley ever reached any conclusions as to the whereabouts of the gold, he kept these to himself.

As bad luck would have it, shortly after Halley had began work at the *Guynie* wreck site he was informed that his old Oxford tutor Edward Bernard was (finally) retiring from the Savilian chair of Astronomy at Oxford. On 22nd June Halley sent a preliminary report on the wreck to his employer Abraham Hill, also begged him *'to intercede for me with the Archbishop Dr. Tillotson to defer the election for a short time, 'till I have done here, if it be but a fortnight: but it must be done with expedition lest it be too late to speak.'* [3] John Tillotson was Archbishop of Canterbury and president of the Savilian electoral board. In the event Bernard did not retire until November and the election did not take place until December, by which time Flamsteed (with a little help from a selfish Newton) had scuppered Halley's chances.

Newton wrote an unusually polite letter to Flamsteed dated 20th August 1691, which was delivered by hand by David Gregory, ex professor of mathematics at Edinburgh University, close associate of Newton and his candidate for the post of the Savilian professorship. Almost everyone else favoured the only other contender, Halley. In his letter, Newton attempted to persuade the Astronomer Royal to publish data in his possession and to undertake to supply him with observations of Jupiter and Saturn for the following 12 or 15 years adding *'If you and I live not long enough, Mr. Gregory and Mr. Halley are young men.'* [4]

What Flamsteed thought of this proposal at the time went un-recorded but Newton's suggestion that Halley of all people might one day carry on Flamsteed's work was an unfortunate blunder or more likely a subtle move. Either way the result was the same; Flamsteed at once set about promoting Gregory's case. He worked feverishly behind the scenes lobbying his religious cronies and spreading malicious gossip about Halley once again. Inevitably Halley was then questioned extremely closely about unfounded allegations that he was a *'sceptic and banter of religion'* by the Bishop of Worcester, Edward Stillingfleet. Halley is said to have replied *' My Lord this is not the business I came about. I declare myself a Christian and hope to be treated as such.'* [5] He was not and Gregory was given the post just as Newton had intended. After all, he already had Halley in his pocket and now had a grateful Gregory's expert mathematical support as well. Halley remained silent but friends cried 'foul', blaming Flamsteed. This was the third time Flamsteed had lied to Halley's detriment.

When he finally got round to replying to Newton's request for data after Gregory had been elected, and in an effort to quell the rumours surrounding the part he had played in the election, Flamsteed's spite got the better of him again. At the beginning of March 1692, he wrote a very long letter to Newton, and sent copies to Wren and others.

He started by trying to explain why he was still not yet ready (after 16 years of observations) to publish *any* of his data, once again using Halley's 'hasty' St. Helena star chart publication as a reason for not falling into the same trap. While he was at it, he attempted to drive a wedge between Newton and Halley by further denigration. In referring to Halley, he wrote:- *'...that I will not be beholden to him for his assistance or advice: that if he wants employment for his time, he may go on with his sea projects, or square the superficies of cylindrical ungulas* [a specific cylindrical section]*: find reasons for the change of the* [magnetic] *variation, or give us a true account of all his St. Helena exploits; and that he had better do it, than buffoone those to the Society to whom he has been more obliged than he dare acknowledge: that he has more of mine in his hands already, than he will either own or restore; and that I have no esteem of a man who has lost his reputation, both for skill, candour, and ingenuity, by silly tricks, ingratitude, and foolish prate: and that I value not all, or any of the shame of him and his infidel companions; being very well satisfied that if Xt* [Christ] *and his apostles were to walk again upon earth, they would not* *escape free from the calumnies of their venomous tongues.'* [6] Within a year of Falmsteed's 3rd libel he had distributed this 4th.

As for the comment '*that he has more of mine in his hands already, than he will either own or restore'* ; in Gregory's personal copy of *Principia* is a note that mentions that Newton had told him that Flamsteed had published as his own some of Halley's lunar data, confirmed because the originals were in Halley's handwriting [7].

Flamsteed had always argued, as he had done in his letter to Moore (chapter 7), that because Halley's St. Helena data had been based on the erroneous Tychonic positions of stars, they were flawed. Halley, who had readily admitted his data were based on these suspect positions, had pointed out that he had no choice as at that time these framework positions were the only published ones. He had to lock his new data into something, otherwise it was meaningless. Although it was tied into Brahe's framework, it could be readily updated if and when Brahe's stars were repositioned and those details published by Flamsteed; who repeatedly refused to do so.

Years later when Flamsteed was preparing plans for the publication of his *Historia coelestis*, Halley's St. Helena material was among the data he intended to include and when his own posthumous publication eventually saw the light of day in 1725 not only did it include Halley's St. Helena data, but there was no mention of who had made those observations.

Because Flamsteed had sent copies of his libellous Newton letter to other prominent astronomers, Halley was left with little option but to refute the allegations; which he did a few months later in the politest possible way.

The RS Journal Book for June 11th 1692 states that *'Halley read a Paper being a vindication of his Observations made at St. Helena from some groundless Exceptions of Mr. Flamsteed's, because of a difference found between his Observations, and those of one Pere Thomas made at Siam lately published by Pere Gony at Paris. He shewed the near agreement of his St. Helena Observations with those of Mr. Richter made at Cayenne about the year 1672, and that the principall differences between Pere Thomas and him fall out where he and Richter do perfectly agree; And Mr. Flamsteed being a Member of the Society he desired the Society's leave publiquely to vindicate himself in print from this aspersion, which was permitted him: All personall reflexion being to bee forborn.* [8]

Halley's reference to Richter's confirmatory observations related to data collected by that highly regarded French astronomer, and Halley clearly stated that he had no wish to become involved in a private war with Flamsteed. The Astronomer Royal had failed to divide the opposition and simply drawn attention to his own deficiencies.

Commentators have questioned Newton's ethics in promoting Gregory's cause at the expense of Halley, the man responsible for publishing his *'Principia'* and adopting the hassle and financial risks that went with it. Certainly however highly Halley rated Newton as a scientist, there can be no doubt his opinion of Newton as a man must have been seriously undermined by the Gregory affair and he surely would have begun to have mild second thoughts regarding the glowing Latin ode to the great man he had written and inserted into the preface of *'Principia'* .

Whatever the reasons for Newton's behaviour, had Halley been appointed to the Savilian chair of Astronomy he might not have gone on his voyages of discovery and Gregory may not have published his own *'Principles of Physical and Geometrical Astronomy',* an important work which went a long way to underline Newton's own research. Halley and Gregory remained friends and eventually Flamsteed came to dislike Gregory nearly as much as he disliked Halley.

After Flamsteed's father had died and left him a small sum, he commissioned a second 140° mural arc, which was more robust than his former flimsy affair. He had the brass arc divided and diagonal scale markings cut (figure 4) (see also chapter 30 and figure 13) by his new assistant Abraham Sharp; in all the instrument costing him about £120 (£24,000). After carefully setting it up on the old north-south wall and adding some 120 pounds of iron bar to the structure to maintain its rigidity he had begun making observations in the autumn of 1689.

Some three years later Flamsteed was able to properly confirm a terrible discovery. Shortly after he had written that libellous letter to Newton but before Halley's stinging rebuttal, he had established that early sightings taken with his expensive new mural arc did not match his latest observations; *'..the distances of the stars from the vertex in the southern quadrant of the meridian, 1'0" or 1'5" bigger than they ought to be.'* [9]

A careful survey revealed that the wall, built on the cheap by Hooke all those years ago, was sinking at the north end, possibly caused by the great and uneven weight of his new instrument. He was left with little choice but to assume that the very slow subsidence was a regular feature and to allow for this as best he could; which he continued to do for the rest of his life. By 1715, the displacement was forcing Flamsteed to apply 14' 20" (nearly half the Moon's apparent diameter) to all observations, a fact he was careful not to advertise.

This unfortunate structural defect was later to be one of the reasons he was so unhappy about being forced to publish this suspect data. Meantime the few assurances he had grudgingly given to Newton would have to be evaded and he was in no position to continue his dispute with Halley either.

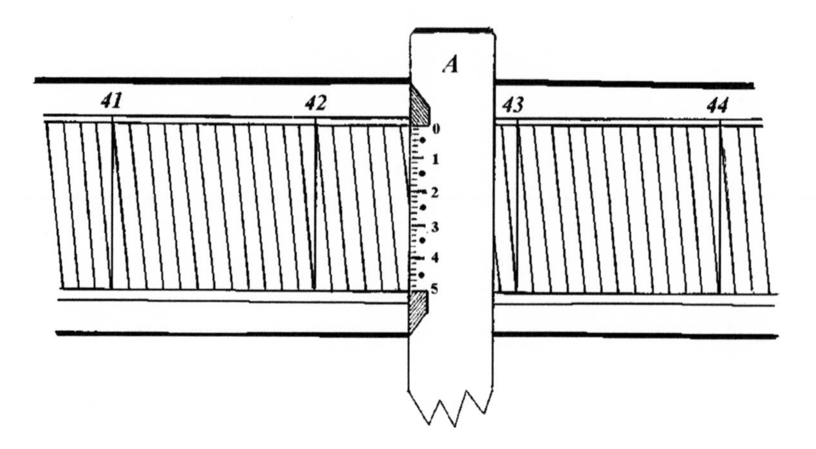

Figure 4. Diagonal scale divisions on Flamsteed's 140° mural arc.
In one version of the diagonal scale, each full degree on the curved brass limb is divided into 5 minutes of arc sections, each marked by a diagonal line (12 per degree). Each of the 30 divisions on the alidade or index arm represent 10 seconds of arc on a graduated scale. The angle represented is 42° 27' 20".

Flamsteed was not the only one to find himself in trouble and in September 1693 Newton suffered a second mental breakdown. Basically this latest collapse which lasted for over a year seems to have been caused by the break-up of his very close but fraught relationship with another of his acolytes Nicholas Fatio de Duillier after the latter's refusal to share Newton's lodgings at Cambridge.

But the waning of Newton's enthusiasm for his cloistered existence coupled with his frustration at his failure to obtain sufficient lunar data from Flamsteed, must all have added to his worries. Yet as with his first breakdown, the problems would miraculously be resolved and Newton would find himself once again in a far better position to press forward with his mission than he had been in before this latest setback.

(1) Eugene MacPike, *Correspondence & Papers of Edmond Halley* (Oxford, 1932), pp.144-5.
(2) Ibid., p.184. (3) Ibid., p.88.
(4) Francis Baily, *An Account of the Revd. John Flamsteed* (London, 1835/1966), p.129.
(5) Colin Ronan, *Edmond Halley; Genius in Eclipse* (London, 1970), p.119.
(6) F. Baily. p.131.
(7) C. Ronan. p.126.
(8) E. MacPike. p.229.
(9) F. Baily. p.55.

CHAPTER 12

Whether or not the *Guynie* incident was a setback for the Royal African Company's new policy of owning vessels, the *Falconberg* purchase was a money-spinner. This armed frigate bought as a replacement for the *Guynie* and named after a major RAC stockholder Thomas Belasyse, Viscount Falconberg, was destined to become the single most successful slaver of all time, despite the steady demise of her owners.

The maiden voyage of this 320-ton vessel, which left the London docks in October 1691, was not particularly noteworthy. Her total journey time for the triangular Atlantic voyage was slow, the number of slaves delivered to Barbados went unrecorded and her return 'third leg' cargo consisted of a mere 19 tons of sugar. It was almost certainly at this point that her captain was replaced.

Falconberg left England outward-bound on her second triangular Atlantic passage on 29th May 1694. She probably passed within sight of the *Guynie* wreck site as her crew pressed on all sail in order to clear the confines of the English Channel speedily, prudently not venturing too close to the French coast. Her cargo would have included iron ingots, many bales of cloth and crates of trading goods. Her unenviable destination on that first leg was the slave market at Ardra on the West African Gold Coast; local RAC employees posted to this hellhole had a life expectancy on arrival of only five years!

Falconberg did not take aboard her new cargo of slaves until the hurricane season was over and so completed the middle passage across to Barbados late in the year. There she unloaded no less than 592 live Negro slaves! This from a ship of about 30 metres in length and not much bigger than a modern luxury yacht. *Falconberg* now carrying 148 tons of sugar worth nearly £300,000 (£60 million), an unspecified amount of private imports and her normal wartime crew of about 60, was back in England by June 1695 having completed the entire three-way trans-Atlantic voyage in a little over a year.

Considering the delays caused by the hurricane season and the need clean out the holds and careen the hull in the West Indies, the captain of this particular ship was either a risk-taker or an exceptionally fine seaman. Because of these unavoidable delays, merchants usually expected their ships to take up to two years for the round trip, as indeed had been the case on *Falconberg's* maiden voyage.

Although slaver captains were rapidly acquiring the expertise to enable them to more efficiently ply their obnoxious trade, the English Royal Navy was still intent on living in the past. When the fleet had eventually put to sea in readiness for the 1691 season of warfare against the French in May, two months after the *Guynie* sinking, they had run into heavy weather and promptly returned to port. In September, having played cat and mouse with the main French fleet for several months, the French Atlantic squadron sensibly retired for the winter to their fortified anchorages at Brest.

The English were left dithering in the western approaches to the English Channel whilst deciding whether to attack Brest or go home, when a rising south-easterly gale eventually forced the issue. Some of the unwieldy English warships, instead of seeking sea room, sought shelter at Plymouth but found themselves trapped on a lee shore and in imminent danger of being dashed onto rocks still short of their safe haven.

The subsequent shambles left *Harwich* wrecked and *Northumberland* and *Royal Oak* aground. Shovell's *London* and several other of the largest men-of-war managed to get their massive and unwieldy anchors to hold at the last minute. Tragically, *Coronation* anchored next to *London*, foundered possibly because of cannon breaking loose, drowning Captain Charles Skelton and all 600 members of her crew.

Following this disaster Shovell was one of the admirals who formally agreed that in future the great battlewagons should, if at all possible, be tucked up in bed for the winter by the end of September; thus limiting their usage to about six months of each year. Advice which he would himself repeatedly ignore. This recommendation would not have applied to smaller vessels and *Falconberg* had sailed through the disaster area on her maiden voyage a month later without her captain giving the matter a second thought.

Shovell already Major of the First Marine Regiment, now wasted no time in asking to be appointed Lieutenant Colonel of the Second Marine Regiment instead. This time the reason for the vacancy was painfully obvious; it had occurred because of the death of the incumbent who had just drowned in the *Coronation* disaster.

Part of Shovell's peculiar plea to Secretary of State Daniel Finch, the Earl of Nottingham read thus:- *'I am humbley to thank your Lordship for the unexpected favour of the being made Major of the first mareen regiment and give your lordship this trouble to begg, if it not be too impertinent, to be Leiut Collonell of the second mareen regiment in the room of leiut collonell Paston who is drowned in the Crownation. I hear my Lord, that tis designed that Rear Admiral Rooke shall have the leiut collenellship;.....'* Shovell went on to explain in typical 'Flamsteed' style why he was far better suited to the post than the aging Rooke and concluded his personal letter by giving his opinion that *'....a major of these regiments ought to be an officer who very well understand the discipling, managing and fighting a regiment ashore, which at present I do not understand. I will at present give your lordship no further trouble.'* [1] Did he really not know how to manage a marine regiment ashore or, having handled the *Guynie* gold incident efficiently from on shore did he now feel it was time to distance himself from the First Marine Regiment? Such was Shovell's influence that although Finch immediately began promoting his case, the king had already signed his commission; Shovell's 5th lucky event.

This latest Royal Navy fiasco during which *Coronation* and *Harwich* had apparently both fallen apart prompted Halley to give some thought to a better method of securing a man-of-war's massive guns in bad weather. He was convinced that upwards of 120 tons of metal on either side of the ship above the water line could all too often batter the entire ship to pieces.

On 24th December 1692 he read a paper to the Royal Society which suggested a system of shifting each gun's securing tackle to an anchoring point on the opposite hull wall, which would remove some of the stress; *'A Method of Enabling a Ship to Carry its Guns in Bad Weather'* [2] . As usual Hooke was moved to intervene. *'Corrected Hallys mistake of.... Hallys way to layh guns absurd.'* [3] He may have been right, but certainly something had to be done to stop these great leviathans from literally falling apart in heavy seas or battering themselves to pieces when stranded on shoals.

Two years later Shovell, having continued in the close company of the monarchy on all possible occasions, was appointed to the lucrative post of Commander of the Navy, which carried an annual salary of no less than £1,000 (£200,000). That same year, 1693 witnessed another navigational blunder involving Shovell.

The fleet was engaged to protect the outward-bound Smyrna merchant fleet of some 400 Dutch and English ships. The planned route was to skirt the Bay of Biscay, then head south to Gibraltar and on east through the Mediterranean to the Turkish port of Smyrna where, as usual they would exchange their manufactured goods for silks, spices and perfumes that had arrived overland from the Orient.

Although a small squadron had managed to rendezvous with the convoy, the main sections of the Royal Navy's escorting fleet, under the joint command of Shovell and two other equally incompetent admirals, was so slow in putting to sea that when they arrived at the assembly point there was not a soul in sight.

When they did catch up with the escort squadron they promptly lost them again in the dark and decided that rather than try to re-establish contact they could safely turn back.

The convoy was certainly now well south of the French naval base at Brest and was thus safe; or so the reasoning went. This odd conclusion allowed the French Atlantic squadron to destroy or capture a considerable portion of the Smyrna convoy off the coast of Portugal two weeks later. Four of the Royal Naval protecting men-of-war from the small escort squadron were sunk with considerable loss of life.

Shovell and the other two admirals sharing joint command escaped the full wrath of a parliamentary enquiry by the closest of margins. Quite how the Commander of the Navy could have been adjudged to have been innocent when he clearly abandoned his escort duties is difficult to comprehend but there was a convenient scapegoat. It would seem that the Second Secretary of State, Sir John Trenchard had been aware that the French Brest-based fleet had sailed but had failed to inform the admirals, so the brunt of the blame was neatly shifted to his plate.

The fact that Shovell should have sent a frigate to check the Brest anchorage but failed to do so was overlooked. Evidence presented by the skipper of a small vessel that had been illegally trading with the French and who had informed all three admirals that he and his mate had actually seen 19 French men-of-war leaving the Brest anchorage weeks earlier was dismissed as hearsay despite being given under oath.

The excuse given before parliament for turning back was that the convoy was thought to be out of danger. In truth there was probably little point in continuing because Shovell's squadrons were incapable of navigating well enough to find their mislaid charges; the techniques employed by the Royal Navy not having improved since the second Dutch war three decades earlier. The saddest part of this latest sorry affair was that the merchant navy had obviously made solid progress. Their compasses were properly serviced, they could determine *latitude* to within about 15 miles and most certainly had the captain of *Falconberg* been advising Shovell, the main fleet could have located the convoy again.

The king, who obviously saw through the parliamentary cover-up, personally banned Admiral Henry Killigrew and Admiral Ralph Duval from the Admiralty Board, stripped them of their rights as Commissioners of the Admiralty and, as an afterthought, discharged them from all military and civil employment. However, Shovell, who quite clearly was equally to blame somehow escaped the wrath of his king; could this oversight have had anything to do with the *Guynie* gold incident or was it merely politics? Yet again Shovell's luck had held.

Gold certainly played a part in Shovell's 7th lucky escape. Because of the Smyrna fiasco, Shovell was temporarily out of favour and Admiral Sir Francis Wheeler replaced him in command of the escort of an early season merchant convoy heading for the Mediterranean in February 1694 in order to sail past Brest before the French emerged from hibernation. Wheeler's warship, the 80-gun first-rate *Sussex,* was carrying 10 tons of gold and 10 ingots of silver intended to persuade the Duke of Savoy to side with England against the French.

A day short of Gibraltar the fleet of 12 ships ran into a storm and *Sussex* capsized or maybe just fell apart. All but two of the crew were drowned and Wheeler's body was washed ashore several days later.

Part of the wreck of the *Sussex* has recently been located and the British government has given permission for a salvage team to search for treasure for its 'archaeological value'. One can only hope that they have better luck than Halley did with *Guynie.*

In April 1696, Shovell was appointed Admiral of the Blue and was promoted from Lieutenant Colonel to Colonel of the Second Marine Regiment with an appropriate salary increase. By the time that the long war with France was concluded in 1698, he had become Commodore in Chief of the Thames and Medway and carried his Admiral's pennant in the 70-gun *Swiftsure.*

The Commodore's character was obviously complex. Outstandingly brave whenever there was so much as a sniff of prize money on the salt air, yet usually sensibly cautious otherwise, although occasionally given to dangerous bouts of stupidity. A good tactician but poorly versed in navigation. A good commander of men and genuinely concerned for their welfare so long as it did not interfere with his ambitions. Brilliant, stupid, fat and greedy are the four words that probably sum up Shovell best; and as one contemporary commented, *'his covetousness knows no bounds.'*

Although the two men were still to meet, Shovell was now obviously held in low esteem by Halley. This latest Smyrna fiasco had cost Halley and his wife's extended family dear. Stockholders or sub-stockholders in the Levant Company, they had all lost money when their under-insured cargoes had been destroyed or captured by the French following the disgraceful behaviour of Shovell and his fellow incompetents. As Hooke was to note somewhat inaccurately in his diary for the 3rd August 1693, *' 2 east india ships sayd to be taken by French in India: not clear - Hallys trade taken by French.'* [3] In reality 40 merchantmen, including four of the valuable Smyrna ships, had been captured and a further 50 destroyed.

(1) Simon Harris, *Sir Cloudesley Shovell; Stuart Admiral* (Spellmount, 2001), p.139.
(2) Eugene MacPike, *Correspondence & Papers of Edmond Halley* (Oxford, 1932), p.164. (3) Ibid., p.186.

CHAPTER 13

At a very early stage in his career Flamsteed had apparently made improvements to the back-staff (figure 1, chapter 2) by replacing the small hole in the shadow vane with a lens. In hazy weather the lens focused the light from the Sun, giving a better result. This caused ructions because Hooke claimed that he was the inventor and had showed the instrument to the Royal Society *'...some years before the* [1665] *Sickness but was not made use of 'till about ten Years after, and yet now meets with general approbation, and is of continual use, and pretended to be the invention of another, tho' my shewing thereof was Printed in the History of the Royal Society.'* [1] And to the confusion of his detractors, he was right! He did not name the thief but as this version of the back-staff used by Halley during his observations on St. Helena in 1677 [2] was named the 'Flamsteed glass' he had no need to.

Since then no English inventor had seriously dabbled in marine angle-measuring instrument design until Halley had a brainwave during the many boring hours spent off the Sussex coast in 1691 supervising his divers. Before diving operations recommenced, Halley read a paper describing his idea for a device made from some strips of wood, a square brass telescope, a lens and a reflector at a meeting of the RS in London on 2nd April 1692 [3] and a sketch was placed in the society's Register Book.

Hooke now claimed he had *'long since invented such an Instrument as this, that he made the same Object glass serve for both Objects'.* He was being a little economical with truth because he was referring to the instrument depicted in figure 2 (chapter 3) which he then produced from the RS repository [3]. There were differences between Hooke's device and Halley's proposal, which were to later prove significant. Hooke's angled polished metal mirror (a speculum) permitted a clear view of the horizon across its *top.* Halley on the other hand was proposing *either* a speculum *or* a piece of glass with foil behind the top half, which permitted a clear view of the horizon *below* it but *through* the clear glass.

Halley quietly gave the matter further thought and then read a second paper to fellows on 26th November 1692 [4] when he diplomatically acknowledged that *'Dr Hook had gone before me'* and announced that he had abandoned the speculum/foil idea in favour of a simple piece of glass roughened on the back side of the top section. *'Having taken of the polish of the back side* [of the top section] *to hinder the double reflection'* [5] of the heavenly object, Halley had realised that he could then see both the image of the heavenly object (reflected off) and the actual horizon (seen through) the same piece of glass quite clearly. Halley had transformed Hooke's simple metal mirror idea into what could become a very useful tool. Anyone looking out through the window of a lamp-lit room towards a garden can see both the garden and the super-imposed lamp's reflection, but in the field of angle-measuring instruments this was an entirely new concept. There was no need to manufacture and polish a speculum as Hooke had done or as Newton had been forced to do for the concave lens in his reflector telescope. However, the instrument Halley described in his paper was still only at best a rough wooden test rig (figure 5) and although he planned to have a brass version made, nothing more was ever heard of this invention. In reality the flimsy instruments of Hooke and Halley would have been little more accurate (and far more difficult to use and even more easily subject to misalignment) than the back-staff, with or without lens, in general use at the time. So although the novel idea of using a single piece of glass to view two images simultaneously was never going to work well with his own angle-measuring device, Halley may well have been responsible for stimulating further research on similar lines (chapters 15 and 29).

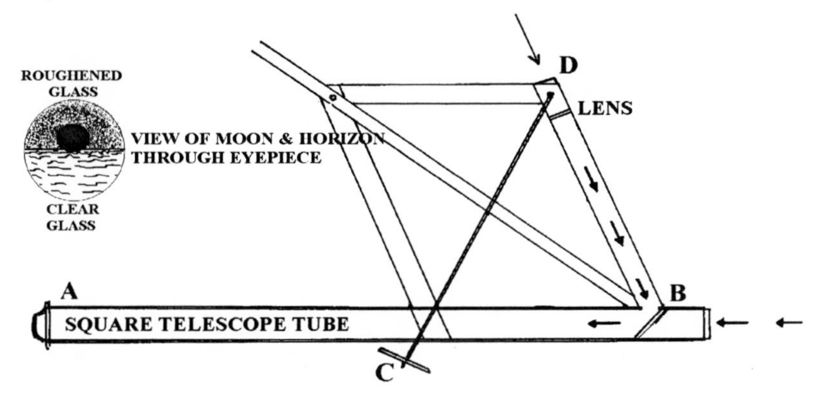

Figure 5. Halley's mirrored angle-measuring device.
The horizon was to be viewed directly through the telescope tube and through the lower section of the pivoted piece of glass at point B. The heavenly object's image would be projected down onto the upper back-roughened section of this same piece of glass through a gap cut into the square telescope tube and on into the eyepiece at A. The Moon or Sun would have been behind the viewer. The device was an advance on Hooke's design (figure 2) because a piece of half-roughened glass replaced the metal mirror which was positioned to allow the horizon to be viewed through *the clear lower section. When the two images just appeared to touch at the join between the rough and clear glass, the angle was read off a scale on the threaded adjuster screw C-D. The view through the eyepiece would be inverted.*

Back in November 1690 Halley had read a paper to the RS concerning *'a Method of makeing observation for determining the Longitude: by placing a Circle accurately turned in the focus of the glass, and noting the entrance, and exit of the Moon and a Star whereby the Difference of the Rt. Ascention of the Moon and star are exactly determined.'* [6] *'Turned in the focus of the glass'* describes a cross-hair placed between the two lenses in the tube of a telescope at the focal point.

The hair would have to be mounted on a ring and carefully lined up vertically to provide a meridian line. The use of cross-hairs (reticules in telescopic sights) in *fixed* telescopes was not a new idea and had been used by astronomers from the early 1640's onwards [7]. What was new was the suggestion that a navigator in the field might use this method to aid in the determination of longitude by the lunar appulse method.

In December 1692, soon after announcing the details of his angle-measuring device, Halley read a second paper on telescopic inserts, which was of even more importance than his first. *'a Method of observing the Moons place by the difference of her Rt. ascension from the fixt starrs: Those he proposed might be obtained every cleare night, and might be a very proper and practicable Medium for finding the Longitudes of places: He shewed likewise how by a Diagonall thread placed at an angle of 45 gr. with the Meridian, the difference of times, between the transit of the said diagonall and the upright thread in the Focus of a Telescope, would accurately measure the difference of declination of any two objects passing through the Telescope. In order to this purpose he produced a Table he had calculated, exhibiting the points of Intersection, and the angle of the Moons way with the Equator for the ready finding at all times, when the Moon would be in the same parallell of declination with any starr proposed.'* [8]

(1) Jim Bennett in *London's Leonardo* (Oxford, 2003), p.74.
(2) Colin Ronan, *Edmond Halley; Genius in Eclipse* (London, 1970), p.35.
(3) Eugene MacPike, *Correspondence & Papers of Edmond Halley* (Oxford, 1932), p.161.
(4) Ibid., p.162. (5) Ibid., p.163. (6) Ibid., p.220.
(7) Allan Chapman, *Dividing the Circle* (London, 1990), p.36.
(8) E. MacPike. p.231.

CHAPTER 14

In the summer of 1694 Newton had more or less recovered his mental health and had concluded that the cause of his breakdown probably had far more to do with the Astronomer Royal than with Fatio. He was convinced that it was solely Flamsteed's intransigence that was preventing him from predicting the lunar orbit. He knew in his own mind that his formula was right but also knew that those very few who actually understood *Principia* were expecting him to succeed or hoping he would fail in the attempt.

However, whereas he had been able to bemuse Flamsteed by predicting that Saturn's orbit was not as the Astronomer Royal had supposed without even putting his eye to a telescope, the Moon was an entirely different matter. Only with more data could he produce the lunar almanac, update *Principia* in spectacular fashion and complete his mission on behalf of his Maker. Then a practical demonstration of the lunar distance method of determining longitude using his tables would advertise his success to the world.

Newton began with an unusual personal diplomatic mission to Greenwich on 11th September. This was partially successful, in as much as Flamsteed supplied him with some 50 or 150 (depending on which of the pair one can believe) lunar observations. Aware of the reason for Newton's re-kindled superficial friendship, Flamsteed also informed Newton that he did not expect him to show this data to third parties or to make use of it without full acknowledgement; both quite reasonable requests.

A month later on 17th October 1694, Newton wrote to Flamsteed requesting lunar data for the months of March, June, September and December going back six or seven years claiming that if Flamsteed could supply these he might well *'set right the moon's theory this winter'*.[1]

Flamsteed had no way of knowing if this was possible but as he did not have the data Newton was unlikely to be able to solve one of the biggest astronomical puzzles of all time *'this winter'*; or could he? If Newton were to succeed, lunar almanacs could be produced and someone else would have solved the longitude at sea problem; he would have to ensure his contribution was properly acknowledged.

Then Newton informed Flamsteed that he had discussed the eccentricities of the lunar orbit with Halley. Flamsteed's worst fears were confirmed; Newton was obviously planning to use Halley's lunar data as well as his (of course he was and why not?) and was trying to use this to put pressure on him. Supply the data I am requesting or I will solve the problem without your help this winter. This would be an absolute disaster for him if he did not comply and an ordinary disaster if he did. He would have to pray that Newton was boasting and help his own cause by prevaricating.

The outlook did not improve when Newton again wrote requesting the lunar data Flamsteed had previously neglected to supply (because he did not have them), modifying his request somewhat and even asking politely after his health. He then pointed out some errors in Flamsteed's computations and asked him to make further observations for the whole of the current month of December 1694.

Flamsteed replied that he was well aware of how important these observations would be because *'I know very well the equations of the moon's motion are the highest this month and next, that they can be again this next 9 years'.* [2] He then casually mentioned that there would be a break in these vital observations because he was taking his Christmas vacation! Flamsteed managed a mere three lunar observations that December and seven in January, all of which were forwarded to Newton.

On another occasion Flamsteed testily noted that *'I was ill now of the head-ache; not being able to calculate, I sent him* [Newton] *the observations that he might compare the moon's places for himself. My work of the fixed stars was interrupted by my distemper.'* [3] Now Newton was intruding on his hidden agenda; no wonder he was suffering from migraine and his gall stones were playing up again.

It later transpired that Flamsteed had apparently taken only 12 lunar observations in the year 1692, 17 during 1693 and 45 during 1694 all of which he had made available to Newton [4]. It is easy to imagine the scene. The astronomer trying desperately to avoid disclosing the paucity of data and his failure to stick with his remit and the mathematician trying to extract every last drop of information without providing any excuse to end their relationship whilst doing so. Flamsteed now desperately needed a scapegoat before his world fell apart and who better than Halley again who had at this point in the story just succeeded in effortlessly embarrassing him over the Tycho Brahe affair in front of Royal Society fellows (chapter 11).

Late the previous autumn Halley had somehow obtained the *'New and Exact Tables for the Eclipses of the First Satellite of Jupiter'* from his good friend Cassini, a no mean feat as England and France were once again in the throes of a major war; the French fleet having just decimated the Smyrna convoy.

In November, the RS had instructed him to convert these to the Greenwich meridian and to publish the tables, which he did, rightly using only Cassini's name [5].

These new tables included all the mathematical information required to compute future observations relating to Jupiter and Io in much the same way as some modern almanacs do. The Astronomer Royal was not amused; if he had not been so pre-occupied he could have maintained his links with Cassini and published the tables himself, or so he believed; now Halley had barged in again. Flamsteed immediately tried to turn this latest embarrassment to his advantage by accusing Halley of all manner of imagined insults.

A good example of Flamsteed's latest unethical behaviour is contained in a letter to Newton written in January 1695 shortly after Halley's publication of the new Cassini almanac. *'I never found any thing so considerable in him* [Halley] *as his craft and forehead, his art of filching from other people, and making their works his own; as I could give instances, but that I am resolved to have nothing to do with him, for peace sake.'* [6]

This 5th unwarranted attack was clearly intended to remove Halley and his lunar data from the equation. If the puzzle was to be solved it would be a Flamsteed/Newton success. But Newton realised he had bled Flamsteed dry and his pestering ceased; he was beginning to realise that he was not going to solve his puzzle so easily after all. A more devious and better planned approach was called for and, rather than fall back into yet another depression, Newton set about its construction with considerable enthusiasm.

(1) Francis Baily, *An Account of the Revd. John Flamsteed* (London, 1835/1966), p.133.
(2) Ibid., p.144. (3) Ibid., p.143. (4) Ibid., p.714.
(5) Monsieur Cassini, *His New & Exact Tables for the Eclipses of the First Satellite of Jupiter* (R.S. Phil.Trans., London, 1694), Vol. 18, pp.237-256.
(6) F. Baily. p.150.

CHAPTER 15

In the years before his latest breakdown, friends had attempted to obtain the Comptrollership of the Royal Mint for Newton but without success. Now strings were again pulled and he was officially appointed Warden of the Mint by Royal Society president and Chancellor of the Exchequer Charles Montague later the Earl of Halifax and lover to be of beautiful young Catherine Barton, the daughter of Newton's stepsister Hannah.

Following on from his comments on the navigational expertise of the Royal Navy (chapter 10), the past president of the RS and ex Secretary to the Admiralty Samuel Pepys was, like so many others, still deeply interested in oceanic navigation, especially the vexing longitude problem. Soon after his appointment as Warden of the Mint in 1696, Newton received a communication from Pepys urging him *'to discover some method of ascertaining longitude out at sea.'* [1]. Coincidentally a wealthy fellow of the RS had for some while been proposing to circumnavigate the globe with Halley; a voyage in temporary abeyance, but one which would certainly benefit from the use of an efficient marine angle-measuring device. An instrument that Newton and Pepys both knew was essential to the longitude quest.

A man of Newton's enormous talent in the fields of optics and metallurgy would have realised that Halley's single-mirrored prototype (figure 5, chapter 13) could easily be transformed into a reliable tool by making a sturdy brass frame and adding a second mirror. Halley, one of the few persons to enjoy the great man's trust and assigned to the temporary Mint at Chester castle as Deputy Controller to assist Newton in England's great re-coinage enterprise, had by then received a written note and diagram from Newton outlining his wonderful invention. Only after Halley's death more than 40 years later, would Newton's note come to light.

As was the nature of the man, a man still smarting from Hooke's comments on his reflector telescope 20 years earlier, Newton then worked on this expensive and time-consuming project in secret, no doubt incorporating any suggestions he had received from the practical naval navigator Halley, and utilising the tools, equipment and skill of the metal workers and engravers employed at the Royal Mint headquarters at the Tower of London. Producing coins with grained edges to a high enough standard of purity and weight to satisfy both the public and the Treasury was a highly skilled business. Too heavy and the Treasury lost money and too light or impure and the public lost confidence. In the days before grained edges, coin could be officially clipped down to size/weight but now the metal had to be very carefully prepared and the sheets or strips of silver or gold produced to the correct thickness for stamping out the blanks. Newton by all accounts played a leading part in overseeing and improving this process and the Mint was probably the only place in England where a large flat sheet of thick brass free of impurities could be produced and engraved with scale markings to a high degree of accuracy at the end of the 17th century (see also chapter 22).

The problem Newton then faced was simply that he was convinced Hooke would call him out the moment he demonstrated his invention regardless of whether or not there was any justification. This was because Hooke and Wren had published drawings of a surveying instrument using two mirrors in *Animadversions of Hevelius's Machina Coelestis* in 1674. Newton was certainly never again going to place himself willingly in Hooke's firing line. So he wisely said nothing, merely agreeing to lend the latest version of his precious instrument to Halley for testing during the proposed circumnavigation. Only then and only if Halley managed to determine longitude with his device would Newton publicly unveil his invention.

At this point in the angle-measuring device saga, a professional sailor entered the story. Not all navigators in the Royal Navy were inept. Unfortunately, those who had slowly fought their way up through the ranks without help from 'on high' all too often found themselves at odds with officers with upper class connections who treated them and their ideas with undisguised scorn.

One such unfortunate man was Edward Harrison, long-time Lieutenant and self-proclaimed expert navigator who had served in no less than six ships of the Royal Navy and who had also spent time in the merchant service. Widely experienced of both Atlantic and Indian oceans he had very definite ideas on navigational methods.

He rightly advocated the proper use of log-lines to determine ship speed, but was significantly highly critical of the landlubber English experts who were being paid to solve the longitude problem. An intelligent but bitter seaman with a large chip on his shoulder; a dangerous combination if you had the misfortune to be a landlubber expert captaining a ship on which he was the first officer.

In 1694, Harrison had proposed to the RS various methods of finding longitude at sea in a paper, which, in fulfilling his duties as Clerk, Halley had personally read to fellows. The Admiralty had then asked him for his comments on Harrison's proposals, and these had not been favourable (chapter 18). When the anticipated acclaim failed to materialise, Harrison published a book which expanded on his RS paper; *'Idea Longitudinis: Being a brief DEFINITION Of the best known AXIOMS For finding the LONGITUDE or A more rational Discovery thereof, than hath been hereto-for Published.'* Printed in 1696 and dedicated *'To the Right Honourable, the Commissioners for Executing the Office of the Lord High Admiral of England. etc.'* , Harrison sent the Admiralty, the Astronomer Royal and the RS copies.

In the book's preface Harrison criticised the RS for ignoring his original paper. *'I discussed this Art* [of determining longitude] *with some Fellows of the R.S. whom I found too much aiming at their peculiar Advantage; therefore I resolved to appear on the Public stage, in Print.'* Just to rub their noses in the dirt, the copy Harrison sent to the RS was autographed; it is still in the society's possession. Most telling of all, in the preface Harrison wrote:- *'Some of the ordinary Mathematicians may hate to be outdone by a Tarporlin, if they have aught to say against him, its because his Practice and Experience may prove him to be a more Competent Artist in Navigation than themselves;........'* Boastful words that very nearly left him dangling from a yardarm three years later.

But what of the 80 pages of *'Idea Longitudinis'* ? The ideas expounded were nothing new and underlined Harrison's lack of understanding of the true extent the Moon's peculiar motions, to say nothing of his own mathematical limitations. Nothing new at all? Not quite. Apart from sensibly advocating the regular use of azimuth compasses throughout the Royal Navy in order to better determine how to make proper allowance for compass variation, there was one point that Harrison mentioned which was apparently novel. An angle-measuring instrument, which he described in code *(it is the Scioptrick, Disptrick, Catopterick, P --- S, ----K, and P.----H----K)* and which he admitted might be nothing more than an *'Idea of my Brain........'.* Despite this, Harrison was of the opinion that such an instrument could solve the longitude puzzle. This notion of his seems to have relied on the dioptra (disptrick), a flat circular surveying instrument first used by Heron of Alexandria circa 50-60AD, which presumably he intended to link with a scioptic ball of some kind (scioptrick), in order to project an image using a mirror, or mirrors (catoptrick).

In the concluding chapter of his book Harrison suggested that *'this Instrument will seem most wonderful, when the best Description is given of it, and when put in common use; to give a true and perfect Description of its secret properties, like the Loadstone, may puzzle the Oxonian Sophist...'* Here he was clearly and rudely referring to Halley.

(1) John Craig, *Newton at the Mint* (Cambridge, 1946), p.93.

CHAPTER 16

A typically brief two-line note in one of Hooke's private diaries outlined a Halley lunar observation, which demonstrated expertise that would shortly stand him in good stead. *'Sat 10.* [20th December 1692] *The shortest Day. Hally & Colson observed the occultation of the Pliades by the moon at 5pp.'* [1]

Halley had made many similar observations in the past, some with John Colson and Peter Perkins, but the observation mentioned by Hooke was unusual. The Moon was almost full and the Sun had not even set at the start of the occultation. Halley had calculated that the Moon would pass in front of four stars in the prominent Pleiades cluster in the Taurus constellation and had made a mask to fit on his telescope to exactly block out the Moon's very bright light (-11.8 magnitude).

Another entry in the same Hooke diary dated 11th (21st) January 1693 mentions that *'Hally of going in Middleton's ship to discover...'* [2] Although the rest of the sentence is missing Hooke was referring to a proposed expedition to circumnavigate the globe to register magnetic variation and attempt to determine longitude, sponsored by Benjamin Middleton and the Royal Society some three months before the society made the formal announcement.

The need to determine magnetic variation at differing locations was of vital importance in the quest to make the oceans of the world safer, although not as important as solving the longitude puzzle. In 1674 off the coast of Newfoundland the magnetic compass needle had been found to point between 15° and 20° to the west of true north, in Hudson's Bay in 1668 it had pointed 19° 15' west and in 1670 in the Magellan Straits between 14° and 17° *east* Yet in the West Indies and in the eastern Atlantic off the Spanish coast there was little discernible variation, the needle pointing more or less in the direction of the polar stars. Therefore, because most navigators had little idea of how to do a running check to determine local variation for themselves, the helmsman could unwittingly head the ship in the wrong direction if relying on his compass.

In theory an expert navigator could do this check and then link the locations that possessed similar degrees of variation; superimpose these 'isogons' on a map and save the run-of-the-mill navigator having to attempt to check for himself. However, there were snags. The skilled observer and the chart user both had to know their position and the variation was itself varying little by little, year by year.

A tempting but fragile link to the longitude quest came through the knowledge that in certain parts of the southeast Atlantic, variation tended to follow the lines of longitude. The isogons might act as a grid; not a very accurate method of determining longitude but better than nothing. Therefore, registering magnetic variation and determining longitude were both of considerable importance to any mission. Although there were many competent navigators who could carry out the first task, Halley was the only man who could possibly achieve the latter.

Certainly when the expedition was finally ready to depart five years later, Halley was well equipped to determine longitude at sea by the lunar occultation/appulse method using Newton's new invention.

Interestingly Hooke also noted a week after his note regarding the intended voyage (28th January 1693) that *'Flamsteed married, rails at Hally'.* [2] The foolish man, having got wind of Halley's venture could not leave well alone, even on his wedding day! Newton, at the time in the throes of his own emotional confusion would not have registered the significance of Flamsteed's marriage, but Flamsteed's bride had brought with her a considerable dowry, which would eventually make the cantankerous Astronomer Royal even more difficult to deal with.

The story of the oddly named *Paramore* (spelt *Paramour* on some naval documents) and the ship's link with Halley is unique in the annals of the Royal Navy. The vessel was originally intended to be built and fitted out using Middleton funding but somehow it had always been an Admiralty project, possibly because the rich sponsor was the son of a Commissioner of the Navy and because the ship was being built at the Royal Navy's Deptford establishment a mile up-river from Greenwich. Admiralty approval for construction was given in 1693 and *Paramore* was ready for launching by 11th April 1694.

Paramore's design was that of a typical shallow-draft 'Pink'. A workhorse of a ship, cheap to construct but a pig to handle in bad weather or strong crosswinds. Not exactly the ideal ship to circumnavigate the globe in! With a displacement of 89 tons, she was 52 feet (16 metres) in length, had a beam of 18ft. (5½ m.) and drew 9½ft. (3 m.). Three-masted and manned by a crew of 15 to 20, she would at least provide considerably more room for her crew than most naval vessels.

The voyage was delayed because of the ongoing war with France and a visit to England by the Tsar of Russia who was allowed to borrow *Paramore* for several months.

Only after Peter the Great, to the considerable relief of his royal hosts, had rushed off to build St. Petersburg and a Russian navy, and Halley had completed his work at the Chester Mint, did Halley have the opportunity to conduct preliminary sea trials, during which he may well have tested Newton's instrument. These trials required alterations to *Paramore* and it was not until the summer of 1698 that the ship was deemed fit for purpose.

During the long delay the original expedition's plan had been altered; first because Middleton was no longer able to take part, but mainly because of pressure from a reformed pirate by the name of William Dampier. Dampier had circumnavigated the globe in the late 1680's, eventually arriving back in England in 1691.

In 1697, he published a book on his adventures *(A New Voyage around the World)* which he sensibly dedicated to RS president Charles Montague. Having got wind of the RS sponsored expedition, he pressured Montague and the Admiralty to consider him for a major role. Halley's circumnavigation plan was reduced to the more manageable proportions of the Atlantic Ocean and the Admiralty provided Dampier with a larger vessel better suited to a longer voyage into the Indian and Pacific oceans.

Eventually a fifth-rate battleship, the 290-ton *Roebuck* was refitted, armed, manned and provisioned for a 20-month voyage and Dampier departed for points south three months after Halley. His instructions were to determine magnetic variation en route and to investigate what lay to the southeast of the Dutch East Indies with a view to opening up the area to English trade and commerce. Determination of longitude was not within Dampier's remit or ability.

Meanwhile Halley's expedition also became a fully-fledged professional Admiralty venture. *Paramore* was now a Royal Naval vessel and suddenly Edmond Halley found himself, at the age of 41, appointed commander of an armed Royal Naval ship and in sole command of a professional crew of 20 sailors. He was also about to become entangled in yet another Flamsteed-inspired plot to discredit him.

Little more than a week before *Paramore's* departure from Deptford, Flamsteed sent a letter to Halley's friend John Colson. Flamsteed first warned Colson against spreading untrue rumours (rumours that Colson was apparently unaware of until receiving Flamsteed's letter) *'that Mr. Newton had perfected the theory of the moon from Mr. Halley's observations, and had imparted it to him, with leave to publish it; and that Mr. Halley would publish it in a short time.'* [3]

Then, in an underhand manner which was sadly becoming a habit, he went on to relay carefully orchestrated untruths of his own: *'Newton's theory, when perfected, must needs agree with my own observations, since it is built, as he freely owns, upon them and his doctrine of gravitation:'* In dealing with Halley, who would not be able to defend himself owing to his imminent departure, he wrote *'Mr Halley's* [data] *could be of no use to him* [Newton] *because he used the Tychonic places of the fixed stars to rectify and state the moons by...'* [4] This despite Halley's comprehensive rebuttal of that charge (chapter 11).

To suggest that Halley would knowingly supply Newton with faulty data was ludicrous. Equally ludicrous was the claim that Newton's 'doctrine of gravitation' was built on Flamsteed's data. In any case some of Flamsteed's data were suspect as well the man knew. For the Astronomer Royal to write a libellous letter to one who had collaborated with Halley on lunar observations was scandalous. Colson must have seen through his ruse but....

Flamsteed ended his letter to Colson with the words *'I shall be at Garraway's betwixt one and two* [on Friday 24th October]. *If you come down hither* [to the Greenwich Observatory] *in the mean time let it not be on Wednesday* [22nd October], *for I have company that day.'* [4]

Who was Flamsteed, the part-time teacher of navigation techniques to budding naval officers, meeting a week before *Paramore* was due to depart from her berth a mile up-stream, someone he would prefer a friend of Halley's did not meet? Almost certainly Lieutenant Edward Harrison, R.N., author of *'Idea Longitudinis'* and fellow Halley hater.

If he was meeting this self-opinionated navigator, it was because Harrison had somehow managed to get himself appointed First Officer aboard *Paramore* at the last minute. Flamsteed could furnish him with various detailed lunar eclipse data and if Harrison's boasts were anything to go by, the professional naval navigator may well be able to use these to out-perform Halley in establishing longitude by the lunar distance method as well.

Flamsteed could also double-check the accuracy of Harrison's back-staff that was fitted with 'his' lens. All this would help Harrison to best their mutual enemy in the race to be first to determine longitude on the high seas. Flamsteed might yet pluck success from the flames of impending disaster, on one front at least.

How did Halley find himself in this tricky position? Having met his crew, he had discovered to his dismay that they were not exactly the pick of William III's powerful Royal Navy, and quite obviously the Commodore in Chief of the Thames and Medway had not let any of his best sailors be risked on such a harebrained venture.

The choice of boatswain, an outspoken belligerent troublemaker, was particularly worrying.

Just how Shovell or the Admiralty expected Halley to accomplish such a daunting task without the assistance of an experienced commissioned officer to help him control 20 or so seamen has never been explained. If Shovel had been involved in crew selection no doubt he would have been pleased had this caused Halley's expedition to fail. Landlubber scientists had no business posing as Royal Naval commanders; most certainly ones who published articles suggesting that his all-conquering men-of-war might fall apart in bad weather.

However, the behaviour of the boatswain/gunner John Dodson had soon persuaded Halley to petition the Admiralty for just such an experienced officer. Almost before his letter of request had been despatched, a naval lieutenant (Edward Harrison) arrived hotfoot from the Admiralty and Halley without further ado, signed him on.

Clearly Harrison had been waiting in the wings, probably tipped off by Dodson. Halley, according to later testimony did not realise his new First Officer's true identity and subsequent events suggest Harrison made no effort to enlighten him. Such behaviour shows Harrison was obviously hell-bent on making a fool of Halley; there is no other possible explanation.

Equipped with Newton's instrument Halley was about to be given the chance he had been patiently planning for ever since he had first realised the possibilities of determining longitude on the high seas with the Moon's assistance as a young undergraduate over 20 years previously.

Unfortunately this small vessel now had two of the world's most able navigators aboard (Harrison was by far the more experienced), each with the same initials and each thinking he was armed with the latest angle-measuring instrument and lunar data. A recipe for disaster.

Watching the loading of six 6-pounder guns and two smaller swivel guns at the Deptford naval yard (where *Roebuck* was still being fitted out) Halley, though knowing nothing of any Harrison - Flamsteed plot, must surely have been apprehensive.

At least now because of the Dampier expedition he was no longer expected to circumnavigate the globe; but the new instructions to *'stand soe farr into the South, till you discover the Coast of the Terra Incognita....'* (appendix 3) in a vessel designed for inshore waters was asking a bit much even of a man of Halley's exceptional talents.

The only time he had been in command of any vessel had been during his hydrographic surveys of the southern North Sea and the salvage work on the *Guynie* wreck. His only voyages out into the great and forbidding Atlantic Ocean had been as a passenger on a large East Indiaman to and from St. Helena 20 years earlier. However, most interesting of all were his special instructions; *'..to improve the knowledge of the Longitude'* .

Halley was being ordered to sail farther south than any man before him, yet the Admiralty were sufficiently hopeful that he and his little ship would survive that they concluded their instructions with the words *'... and when you return to England, you are to call in at Plymouth and finding no orders there to the contrary, to make the best of your way to the Downes,*[the big naval anchorage in the English Channel guarded by Deal Castle] *and remain there until further Order.'* In other words, Halley was to report to Shovell directly on his safe return. These same Admiralty instructions ordered him to keep a journal of his voyage, an instruction that was surely unnecessary.

More than a year previously, on the 8th March 1697, shortly before the Tsar had handed *Paramore* back to the Admiralty, *Falconberg* had left England for the West African Gold Coast holding fort of Whydah and was back within 12 months with a cargo of 240 tons of sugar, having deposited 502 living Negroes in Barbados en route.

Then on the 3rd August 1698, whilst *Paramore* and *Roebuck* were both still being fitted out at Deptford, the menacing shape of *Falconberg* slid silently past them on the turning tide heading for the Gold Coast once more, the cloying smell of human excrement ingrained into her timbers masked by the stench of the great river's sewage.

(1) Eugene MacPike, *Correspondence & Papers of Edmond Halley* (Oxford, 1932), p.185.
(2) Ibid., p.186
(3) Francis Baily, *An Account of the Revd. John Flamsteed* (London, 1835/1966), p.162.
(4) Ibid., pp.162-3.

CHAPTER 17

*Note. ** in chapters 17,18,19 & 21 link to Appendix 4.*

Many of the astronomical observations made by Halley during the voyages described in the following chapters have been until recently impossible to quantify, but in this computer age using SkyMap © or similar software it is now possible at least to identify most, if not all the stars and the moons of Jupiter that Halley made use of to determine his longitude en-route.

If only he had included all the information in each instance in his journal instead of omitting data he considered unimportant! If he did keep separate records of all his mathematical calculations, calculations that all too often must have been enormously complex and undoubtedly groundbreaking, they have been lost.

Preparations for a voyage *'...to improve the knowledge of the Longitude and variations of the Compasse'* were difficult enough to handle, but when the Admiralty were also expecting Halley to take his little ship *'...... soe farr into the South, till you discover the Coast of the Terra Incognita....'* he wisely decided on some more sea trials before setting off.

The captain had carefully packed away Newton's instrument and double-checked that he had all his astronomical data, charts and pilot books on board. His wife and three children had said their farewells and were back on the quayside.

Mary was trying her best to appear composed but the children were waving excitedly and if they too were wondering when they might see their father next they certainly did not show it. At least Halley was not so pre-occupied with his duties of command that he failed to acknowledge their waves as *Paramore* was towed from her berth out into the Thames. The crew were still hoisting sails as she drifted slowly downstream past Greenwich Park with its little observatory perched atop the hill. One cannot imagine Flamsteed was not tempted to turn a telescope in *Paramore's* direction.

The date was Thursday 30th October 1698 and Halley was concerned about his late season departure. They should have left months earlier, ideally in July or August as *Falconberg* had done. His plan was to bed in ship and crew on a short journey round the North Foreland and through the English Channel to Plymouth where he could sort out any minor problems before heading out into the Atlantic. However, *Paramore* ran into a northerly gale once in the Channel and had to seek shelter in Portland Roads off the Isle of Wight. The ship was handling badly, planks had been poorly caulked and the wet sand ballast had shifted and was clogging the pumps. Here Halley took magnetic bearings of Weymouth harbour using one of the azimuth compasses issued to him by the Admiralty. His results did not tally with the information provided by his pilot book, which came as no surprise.

Finally, the strong winds abated and Halley backtracked to Portsmouth where he was sufficiently aware of Royal Naval protocol to fire a 5-gun salute to Admiral John Benbow; a salute that was formally and promptly returned. The leaks were repaired, the ballast changed for gravel and *Paramore* eventually set sail on 9th December 1698, now conveniently in the protective company of Benbow's battle fleet which was heading for the West Indies via Madeira.

Paramore's surgeon George Alfrey was a personal friend and early on in the voyage, Halley also developed a friendship with his servant Richard Pinfold. But he had already taken a distinct dislike to his fellow officer, Edward Harrison and his boatswain/gunner John Dodson. In Halley's eventful, long and distinguished career, he really only ever expressed dislike for two people, and these were they; the feelings were mutual. The ignorant John Dodson, clearly under Harrison's influence, never did realise that Halley was actually a far better navigator than Harrison; his commander possessed better equipment, better data and a far superior understanding of mathematics and astronomy.

What went through Harrison's mind when he first watched his captain taking sights with Newton's revolutionary new instrument can only be guessed. Did he believe Halley, having read his book, had stolen his nebulous idea of mirrors and lenses and somehow produced this angle-measuring device? Probably.

Following Halley's demonstrably expert use of an azimuth compass in Weymouth harbour and the unveiling of Newton's instrument, Harrison must have realised his sarcastic 'Oxford sophist' published comment could prove a distinct embarrassment if or when Halley discovered his true identity. In any case how was he going to out-perform this man in the race to be first to determine longitude at sea when his back-staff was so obviously inferior? He had to hope that his superior knowledge of the powerful Atlantic currents together with the lunar eclipse data supplied by Flamsteed would give him the upper hand.

During this leg of the voyage Halley insisted that the ship's speed was checked regularly and he compared this with the daily noon latitudinal Sun sights taken with Newton's instrument.

Travelling southwards across latitudes, it was relatively easy to discover a south-flowing current was aiding their passage; not a strong current but something between ¼ and ½ a knot was adding between 6 and 12 miles to their daily 'through the water' dead reckoning/distance run (DR) record. In this instance the speed and direction of the current was known to the fleet navigators so there were no nasty surprises, but a journey in full overcast without knowing of the southward set could result in bumping into one of the Madeiran islands a full day or more before lookouts had been posted. Which is why a full understanding of all the Atlantic Ocean's surface currents was such an obvious priority for the expedition.

Little *Paramore* had no difficulty in keeping station with the fleet simply because the largest men-of-war, some weighing over 1,400 tons and really only capable of sailing down-wind, were marginally speedier than floating bottles unless there was a fair gale of wind at their backs. At an average speed of less than three knots, it took them 16 days to reach Madeira.

Paramore parted company with Admiral Benbow's fleet in Funchal roadstead. The onshore winds made the exposed anchorage dangerous for large men-of-war and Benbow decided to press on. Halley in the small *Paramore* dropped anchor because he wanted to take on some barrels of Madeira wine and to measure the magnetic variation, both of which, with difficulty he achieved. He had hoped to verify the longitude of Funchal by the Jovian moons method from onshore but was unable to do so.

He had arrived several days too late and at night Jupiter was now below the horizon. He did succeed in taking a noon sight of the Sun whilst at anchor which placed Funchal at 32°30'N, (actually 32°38'N, some eight miles in error; as it would turn out the worst result Halley would ever obtain using Newton's instrument).

He also somehow managed to fix Funchal's longitude at 16°45'W, *'by my reckoning',* which was also some eight miles in error. Considering that the current edition of *'The Mariners New Kalendar'* (one of the fleet's pilot books) placed Funchal at 32°25'N., and was well over 100 miles in error at a longitude of 19°05'W, Halley's longitudinal estimate was a remarkable feat of 16 days of DR. A feat only made possible because his daily latitudinal 'fixes' had possessed an accuracy way beyond the capacity of a back-staff, with or without a Hooke/Flamsteed lens.

Harrison fared far worse as later recorded by Flamsteed. *'I could not observe in ye Road by reason of ye Rowling and quick motion no doubt of ship and complained me in defense that by reason of ye smallness of ye ship it rowld and moved so fast yt he could not observe so exactly as they used aboard ye bigger and steadier East India Merchant men.'* [1] Harrison's failure to determine his latitude in Funchal bay confirms that he was trying to use a back-staff because his essential northerly horizon would have been blocked by the town of Funchal with its backdrop of steeply rising mountains.

Equally, this is conclusive evidence that Halley was using a twin-mirrored device that enabled him to face south to view the noon Sun and to view an unobstructed southerly horizon at the same time, as is done with a modern marine sextant.

The fact that he could achieve a reasonable result despite the heavy-weather conditions so graphically described by Harrison, confirms Newton's instrument also possessed other marine sextant attributes (chapter 30).

Flamsteed, being unaware of the quite amazing situation unfolding off the African coast, was so confident of his spy's abilities that a week after *Paramore* had departed from Madeira he included an ill-advised and boastful postscript to a letter to Newton following receipt of new corrected lunar orbital predictions from the great man based on his gravitation mathematics. *'P.S. In your letter, you say, these corrections will answer all my observations within 10 minutes* [of arc]. *Mr Halley boasts that those you have given him will represent them within 2 or 3, or nearer. I wish you many happy years. J.F.'* [2]

Halley, with a degree of good fortune, and by making use of Newton's (and his own) superb mathematical skills would shortly prove the lie to Flamsteed's sarcastic comment almost before the ink was dry.

Harrison on the other hand, having failed dismally to match Halley's navigational prowess must surely have realised that he was now left with little option if he and his sponsor were to be successful. Halley's superior expertise would have to be sabotaged.

Paramore now ploughed a lone furrow south, standing clear of the Spanish-owned Canary Islands and, no longer under Benbow's wing, keeping a weather eye open for pirates.

They took on water in the Cape Verde islands where they were fired on by two English merchantmen who ironically mistook *Paramore* for a corsair. Having given one of the captains a severe dressing down Halley was ready for the big scientific adventure; a diagonal crossing of the Atlantic Ocean down to Brazil, the southern sector of the infamous 'middle passage'.

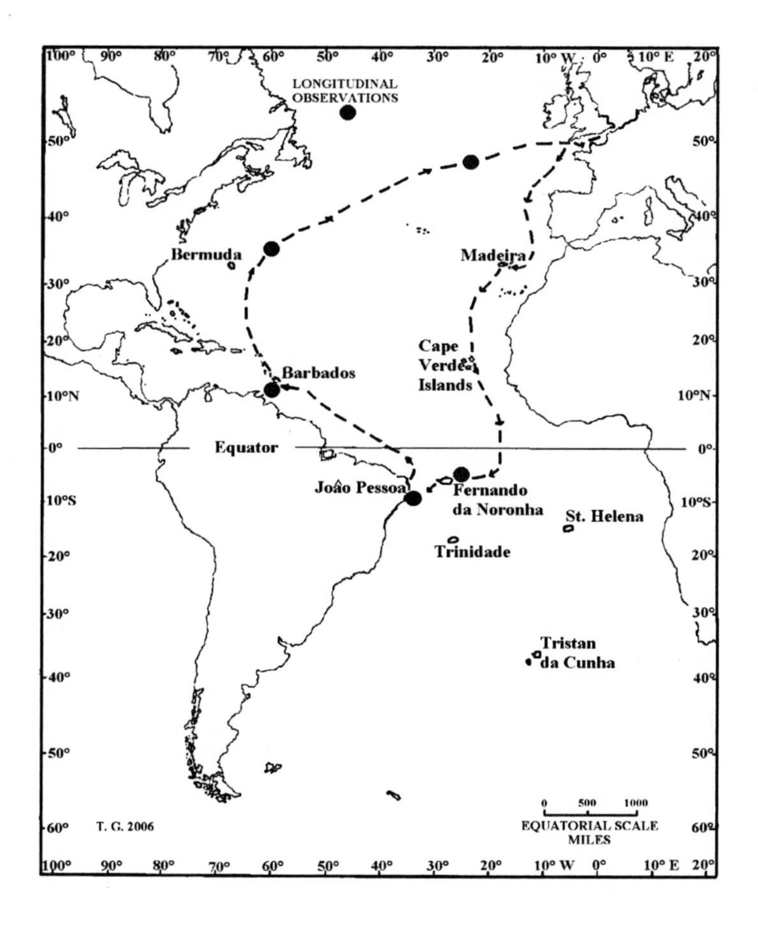

Figure 6. The track of Paramore's first Atlantic voyage.

Now he was ready to check the log-line against the west-flowing current, the big surface current that pushed ships west whether or not they intended to head in that direction. A current every seasoned mariner was aware of but one to be feared because its exact strength was unknown.

Unfortunately, Halley had still not been able to verify his longitude and departed from the Cape Verde island group heading first south and then southwest towards Brazil thinking he was 55 miles to the east of his true position having had to rely on DR ever since leaving England.

Before commencing the trans-Atlantic crossing, Halley called his crew together. He again emphasised the importance of log-line procedures and again pointed out the need to steer compass headings according to his meticulous en-route azimuth compass checks that continually updated the local magnetic variation. No falling asleep at the wheel, no letting the ship stray off course without noting every detail in the log. But he knew in his bones he was going to have trouble with Harrison and Dobson.

Even Dr. Alfrey, who was not a particularly perceptive man, realised that Harrison bore some sort of inexplicable grudge against Halley and he had noticed that the captain was no longer discussing navigation with his First Officer. Indeed, as the weather improved and the heat of the Sun gradually turned skins from white to pink and finally to a healthy tan, it was he, a ship's surgeon who often assisted his captain, writing notes, checking the solar tables and even working on the simpler mathematics.

It was Alfrey, not Harrison who was being instructed in the mysteries of using Halley's peculiar twin-mirrored angle-measuring device. An instrument which was handled like a rare piece of Oriental porcelain and which was normally locked in a large padded box inside the captain's sea chest. But when Halley attempted to explain the mathematics required in order to compute for parallax, index error, refraction and otherwise determine longitude by this lunar distance method, he had to admit that leg amputations presented far fewer problems.

Halley informed his crew that he intended to call in at the uninhabited island of Trinidade, an island which was several days sailing short of the Brazilian coast. He needed to get ashore there with his long telescope and tripod, where he planned to make use of his Halley/Cassini tables to obtain a longitudinal fix using Jupiter's moon Io. The island's position was shown on his chart but the longitude reference was naturally suspect; with a little luck Halley could fix the island's longitude beyond reasonable doubt.

What he probably did not explain to anyone was that there would be several opportunities to determine longitude by lunars shortly before they reached the island. The tried and tested Jupiter method would then hopefully confirm the accuracy of his lunar data and importantly his mathematics. Unfortunately, *Paramore* ran into the doldrums and progress became painfully slow; so slow that they ran short of water and Halley wisely abandoned his objective in favour of the nearer and more northerly Portuguese-owned island of Fernando de Noronha.

On the 17th February 1699, *Paramore* crossed the line; she sailed into the southern hemisphere as Polaris disappeared beneath the northern night-time horizon. On the 24th February, three-quarters of the way across the Atlantic, Halley's patience was at last rewarded. His private log for that day states somewhat economically *'This morning I observed the moon apply to a starr in fascia* [Sagittari sign] *boreali* [the northern band of the scarf] *and concluded myself 160 leagues* [480 miles] *more westerly than our account, and but 50 leagues* [150 miles] *to the East of Fernando Loronha* ' [3]

Following the observation he added *a full eight degrees* of longitude to his DR estimate **. The star involved in this historical high seas longitudinal fix was Rho[1] Sagittari.

Having achieved what he fondly hoped was a very reasonable longitudinal position by the lunar occultation method (the Moon passing in front of a star), the first person ever to manage this highly skilled feat at sea by a margin of 70 years or more, Halley's entire expedition was cut short by Harrison's first serious effort to prove his captain's navigation was dangerously suspect and at the same time attempt to rob Halley of the chance to confirm his recent longitudinal fix.

Any hopes that Harrison may have entertained in being able to show up faults in his captain's knowledge of the South Equatorial current and navigational limitations had been dashed by the Oxford sophist's lunar longitudinal 'fix'. His own extensive hard-won knowledge of this mighty current's true speed would not help him now. Dreams of the captain being smugly informed that the coast of Brazil would be in sight shortly when the ignorant scientist thought it was still 500 miles away had turned into a nightmare the moment Halley added that 8° to the DR position in the ship's log.

The little ship was now being steered westwards along a horizontal band of latitude directly towards Fernando de Noronha, whose latitude was approximately known but whose longitude was uncertain, but the captain's noon Sun sights were now showing the ship was being set to the north; the strong west-running current was now a north-westerly one, which was unexpected and puzzling and did not accord with Halley's sparse pilot notes.

The next night Halley was woken by that sixth sense that so often warns sailors of danger even in the deepest of slumbers. The ship's motion had changed. Peering in the moonlight at the compass he kept by his cot, he realised that the helmsman of the watch, the boatswain Dodson, was not steering as instructed. As Halley angrily wrote in his log '*...my boatswain who had the Watch, Steard a way NW instead of W.. I conclude with a designe to miss the Iseland, and frustrate my Voyage, though they pretended the Candle was out in the Bittacle, and they could not light it.*' [4]

Harrison had persuaded Dodson that Halley's navigation was badly at fault, and had they been permitted to maintain that north-westerly course the entire crew could well have died of thirst. The nearest land in that direction was over 1,000 miles away.

Harrison was probably aware of this and would have brought the ship's heading back on track after Fernando de Noronha had been missed to Halley's embarrassment.

The angry captain took command, reset the course and the following afternoon a speck in the vast southern Atlantic Ocean, the island of Fernando de Noronha hove into sight. But a despondent Halley had realised that the entire expedition had been jeopardised. How often had the pair disobeyed his sailing directions in the past? Were not his explicit instructions from the Admiralty to plot the positions of all lands encountered? How could he achieve this when subordinates chose to alter course to avoid such lands and maybe put their shipmates' lives in danger into the bargain? Yet almost unbelievably Halley still did not realise Harrison was the author of *Idea Longitudinis*.

Unfortunately when Halley, wisely taking Harrison and some of the crew with him, landed on the desolate scrub-covered Fernando de Noronha they could find no water and Halley failed to obtain a longitudinal fix using Jupiter although he never explained in his journal why. Cloud cover, strong winds which shook his long telescope or his unwillingness to remain ashore overnight (Jupiter was not rising until after midnight) in the peculiar circumstances prevailing; all could have been explanations.

So instead of being able to confirm that historical first lunar 'fix' of 24th February he was forced to rely on it. He fixed the longitude of the island as 31°40'W. (23°40'W of DR + the 8°). Halley then fixed the latitude as being 03°57'S by taking a tricky noon Sun observation, the Sun being at an altitude of 86° or almost directly overhead. This incidentally confirms that Newton's instrument was capable of measuring angles as wide as 90°. The true position of Fernando de Noronha is 03°52'S, 32°23'W; Halley was 5 miles in error latitudinally and at most 43 miles in error longitudinally.

Four days of sailing SW took them to the coast of Brazil and there on the very night of his arrival on 7th March, Halley finally obtained the confirmatory longitudinal lunar appulse fix (the moon passing close to a star), he had been so patiently waiting for. *'On the night we fell in with the Coast viz Februr 25th* [7th March] *I observ'd the Moon to apply to the Bulls Eye and that the starr was in a right line with the Moons horns when it was 10 deg 26 min high in the West, or at 10h 11' 44"* [pm estimated local time] *from both which observations I conclude that the Longitude of this Coast is a full 36 deg to the Westward of London wherefore we have been set by the currents to the Westwards, during the long calms, not less than 200 leagues.'* ** [5]

Edmond Halley, navigator extraordinaire and despite the efforts at sabotage, was a single degree in error. His longitude was actually 35°00'W. Somehow, in four days of sailing from Fernando de Noronha his longitudinal error had moved from minus 45' to plus 1°. Either one of the lunar 'fixes' was wrong or the crew were still playing stupid dangerous games; in all likelihood both 'fixes' were probably in error by nearly a degree.

Interestingly this log entry was the first (of two) to mention seconds of time and the first to confirm he was using his previously described cross-hairs insert in his telescope [6]. The very bright star Halley was referring to was Aldebaran in the Hyades cluster of the Taurus constellation. In Halley's day astronomers knew this star as Parilicium but sailors called it the Bull's Eye because of its position in the constellation - the eye of Taurus, the bull.

In taking stock of his position, Halley concluded that the west-flowing current had taken *Paramore* more than 600 miles farther west in 43 days than the log line had indicated. The surface equatorial current was running at over ¾ of a knot.

Halley sailed to the nearby Portuguese port of Joâo Pessoa on the Pariba river estuary and anchored off, waiting for a pilot. However, the Portuguese were rightly suspicious of a lone Pink flying the flag of the English Royal Navy; even if it was lightly armed.

Halley would probably have given his eye teeth (or Harrison's) in return for permission to survey the broad estuary which was no more than a large blank on his chart, but the Portuguese would have none of it. On the 15th of March, Halley yet again failed to obtain a satisfactory longitude fix from an observation of a lunar eclipse owing to cloudy conditions **.

The cloud also put paid to Harrison's independent effort at outdoing Halley and Flamsteed would now not know if his own lunar eclipse mathematics were superior. Harrison then copied the longitude 'fix' achieved by Halley on 7th March into his own log and later managed to give Flamsteed the impression that this was his own result [7].

Eventually the Governor permitted them to take on fresh water and Halley was allowed ashore, although it is not clear if the Governor realised he had taken instruments (but not his tripod) with him. Sadly they were of no use. On 17th March he attempted to confirm his longitudinal position by observing the eclipse of Jupiter's moon Io but failed owing to *'the great height of the Planet, and want of support of my long Telescope.'* [8] So

Halley was forced to be content with that lunar close encounter with 'the Bull's Eye' of 7th March.

It must have been at about this point in the voyage that a disgruntled and depressed Halley decided to head for home. Not to give up, but to have his First Officer court martialled. After some considerable thought and a careful reading of his commission he changed his mind and decided to unload Harrison and Dodson in Barbados *'in order to exchange them if I found a Flagg* [an admiral] *there.'* [9] before resuming his voyage of discovery.

On nearing Barbados, Lieutenant Harrison, now very well aware of his captain's intentions, again deliberately disobeyed orders whilst Halley was sleeping and steered to windward of the island in an attempt to prevent a landfall. If he was going to be court martialled he was going to make sure it was back in England and not in some godforsaken foreign port. In any case, a bit more breathing space might yet allow him to turn the tables on Halley.

But again Harrison had underestimated his captain's seamanship and was again caught out. Halley managed to dock successfully in Barbados but to his dismay there was no flag officer on hand. Whilst ashore Halley observed an eclipse of Jupiter's moon Io** and concluded that if the tables were accurate the longitude of Barbados was 59°(0?)5'W [10]. He had now also accurately, if inadvertently assessed the speed and direction of part of the North Equatorial current which had pushed him north and west towards Barbados from Brazil far quicker than expected.

It would give a nice touch to the story if Halley were to have met up with *Falconberg* in Barbados. Indeed he may have done; *Falconberg* had recently unloaded 471 live slaves there.

Benbow's West Indies squadron was out on the high seas, where exactly no one knew, so Halley headed north again for Bermuda, his last chance to rid himself of his troublemakers. Again, under the influence of Harrison, the ship was repeatedly steered east whenever Halley was not alert. This meant that when *Paramore* reached the latitude of Bermuda the island was nowhere in sight.

This time Harrison had got the better of him but three days later Halley managed to take a lunar observation that confirmed his suspicions, placing *Paramore* well to the east of Bermuda. *'This Evening I observed the Moon to apply to a Starr in the foot of Leo.'* ** [11] The first use of a longitudinal fix to unmask a mutineer.

Although Halley could have headed west and made a landfall in Bermuda he still did not know if he could find a flag officer when he arrived, so clearly still very angry, he decided to head for home.

But Harrison was not finished with his sabotage efforts. As Halley was later to notify Admiralty Secretary Josiah Burchett, in explaining his premature arrival in Plymouth; *'But a further motive to hasten my return was the unreasonable carriage of my Mate and Lieutenant, who, because perhaps I have not the whole Sea Dictionary so perfect as he, has for a long time made it his business to represent me, to the whole Shipps company, as a person wholly unqualified for the command their Lopps have given me, and declaring that he was sent on board here, because their Lopps knew my insufficiency. On the fifth of this month* [15th June 1699] *he was pleased so grosly to affront me, as to tell me before my Officers and Seamen on Deck, and afterwards owned it under his own hand, that I was not only uncapable to take charge of the Pink but even of the Longboat; upon which I desired him to keep his Cabbin for that night, and for the future I would take charge of the Shipp myself, to shew him his mistake.'* [12]

On the 22nd June Halley knew another opportunity would be given him to check his longitude - *'in the Morning the Moon apply'd to a Starr in Line piscium by which I find my Self 25 leagues more westerly than my Reckoning.'* ** [13]

A week later when the Scilly Isles hove in sight Halley obtained a very good noon sight which enabled him to fix the northern limit of the islands *('haveing a very good observation')* as being at a latitude of 49° 57'N. This was only two miles in error.

Flamsteed later wrote [7] that Harrison informed him that he had taken a latitudinal noon sight on 19th June, which accorded reasonably well with Halley, and claimed to have somehow fixed the ship's *longitude* on 25th June. There was absolutely no way in which Harrison could have done this and he must have simply made use of his knowledge of the effects of the Gulf Stream to manufacture a figure, one which he confidently but erroneously expected to be far more accurate than that of his captain.

On 2nd July 1699, *Paramore* was towed the last mile into Plymouth Sound and dropped anchor. It was here Halley sent that dispatch to Burchett. Significantly Halley had also mentioned his success with determining compass variation and advised their Lordships that he had devised a method of determining longitude at sea ...*'Which I have severall times practiced on board with good success'.* [12] He requested his orders into the Downs anchorage be expedited so that he could come to London and give them a fuller account of this incredible news.

On receiving Halley's despatch Burchett went into action. Their Lordships at the Admiralty were informed and instructions were sent to Shovell at the Downs to convene an immediate court martial to try Harrison, enclosing a copy of Halley's letter of complaint, which was to be urgently and strictly enquired into to enable Halley free to report to the Admiralty in London.

Unfortunately *Paramore* arrived at the Downs anchorage before Shovell had received these instructions so, with the permission of Admiral Sir Cloudesley Shovell, Commodore in Chief of the Thames and Medway, Halley took a carriage to London to report directly to Burchett, handing over *Paramore's* log books and providing him with a report of the expedition's results, results obtained with the aid of Newton's revolutionary angle-measuring device.

He managed to spend one night at home recounting his adventures to his wife and wide-eyed children before receiving orders to return at once to the Downs. The captain of *Paramore*, probably accompanied by the Comptroller of the Navy Sir Richard Haddock, (the man who had escaped from the burning *Royal James* by diving through a gun port) was soon bumping his way back to Deal and on the 13th July he found himself aboard *Swiftsure,* Shovell's 70-gun flagship where the court martial was to be held.

Falconberg had followed *Paramore* across the Atlantic, docked during the week following the court hearing and began unloading its main cargo of 228 tons of sugar valued conservatively at £140,000 (£28 million).

(1) E. Forbes, L. Murdin & F. Willmoth, eds., *The Correspondence of John Flamsteed* (Bristol, 1997), Vol. 2, pp.811-813.
(2) Francis Baily, *An Account of the Revd. John Flamsteed* (London, 1835/1966), p.165.
(3) Norman Thrower, ed., *The 3 Voyages of Edmond Halley* (London, 1981), p.98.
(4) Ibid., p.99. (5) Ibid., p.103.
(6) Eugene MacPike, *Correspondence & Papers of Edmond Halley* (Oxford, 1932), pp.220-231.
(7) E. Forbes, L. Murdin & F. Willmoth, eds., Vol. 2, pp.783-4.
(8) N. Thrower, ed. p.102.
(9) Ibid., p.104. (10) Ibid., pp.106-7. (11) Ibid., p.111.(12) Ibid., pp.281-2.
(13) Ibid., p.112.

CHAPTER 18

*Note. ** in chapters 17,18,19 & 21 link to Appendix 4.*

With Sir Cloudesley Shovell presiding and Sir Richard Haddock, Rear Admiral Sir Basil Beaumont and eight captains (including Sir William Jumper and John Price) as members in attendance, this exceptionally high-powered court martial listened to Halley's evidence, such as it was.

Halley had faced a moral dilemma. He knew he had sufficient written evidence to have Harrison convicted of incitement to mutiny. If he pressed such a charge too vigorously, Harrison might possibly face the death penalty and his own chances of completing the unfinished portion of his voyage of exploration might be severely delayed. So he gave evidence only of insolence.

Unfortunately Shovell had decided to charge Harrison with refusal to obey a command, something Harrison had been careful to avoid doing and which Halley had never complained of. Whether this was a plea-bargaining arrangement between Shovell and Harrison, we will never know. Accept a severe rap over the knuckles and leave the service in return for keeping a closed mouth over Newton's instrument or we introduce the serious charge and you hang! Certainly Burchett knew of Harrison's confession (chapter 17).

The Admiralty's orders to Shovell were to try Harrison but the court decided to enquire into the behaviour of the other troublemakers as well. Because of Halley's reluctance to press the serious charge the hearing was, by accident or design, a farce. The court's official report and decision fails even to name any of the accused other than Harrison.

Had any of the 12 experts bothered to pay close attention to the logs or to Halley's journal they would have noticed that he had placed the northern limits of the Scilly Isles at 49°57'N *latitude* and not some 15 miles further north as was noted in *Swiftsure's* latest edition of *'The Mariners New Kalendar'* or any of the other pilot books and charts their ships carried. They might even have noticed the compass variation at their Downs anchorage was given as 7°26'W., a dangerously far cry from the 4° W. most of them imagined.

What then annoyed Halley was the wording of the court's decision. *'Enquiry was made into ye Complaint exhibited by Capt Edmd Halley Commander of his Majties Pink ye Paramour against Mr Edwd Harrison Lt & Mate & other Officers of ye Sd Pink for Misbehaviour & Disrespect towards him their Comander. Upon a Strickt Examination into this matter ye Court is of opinion that Captain +++ Halley has produced nothing to prove yt ye said officers have at any time disobey'd or denyed his Comand tho' there may have been some grumbling among them as there is generally in Small Vessels under such Circumstances, and therefore ye Court does Accquit ye Sd Lt Harrison and the other Officers of his Majties Pink ye Paramore of this Matter giving them a Severe reprimand for ye Same.'* [1]

To add insult to injury, Halley only rather late in the day apparently discovered the true reason for Harrison's enmity as he revealed in a letter to Burchett dated the day after the hearing. *'My Lieutenant has now declared that I had signally disobliged him, in the character I gave their Lopps of his Book, about 4 years since, which therfor, I know to be the cause of all his spight and malice to me, and it was my very hard fortune to have him joyned with me, with this prejudice against me.'* [2]

It is a little difficult to believe that the penny had not dropped earlier. Halley and the Royal Society had both been grossly insulted in Harrison's book, a publication that had clearly stated the author's full name, naval rank and credentials; a copy of which was in their possession. Surely neither Hooke nor Newton would have been so forgetful.

What Halley's reaction would have been had he got wind of Flamsteed's involvement does not bear thinking about. Not only had Flamsteed apparently known of Harrison's intentions, but following the court martial, Harrison supplied him with a copy of his personal log of the voyage, details of which Flamsteed had then carefully kept to himself [3]. This behaviour of the Astronomer Royal was beyond the pale. The RS, of which Flamsteed was a fellow, had co-sponsored the expedition that had been brought to an ignominious end as a direct result of Harrison's behaviour.

Then again, how could the Admiralty have agreed to place Harrison in a key position on such an important expedition? An officer whose theories on the longitude question had been investigated at their request and found wanting by the commander of that very expedition.

The wording of the court's findings had made Shovell's (and presumably Haddock's) opinion abundantly clear. Amateur sailors, however expert, had no business commanding Royal Naval vessels.

Within days they would both be forced by political necessity to swallow their pride, but this bigoted attitude was later to cost Shovel and nearly 2000 sailors their lives. The Admiralty also seemed unable or unwilling to implement Halley's suggestion that they work at ways of making the great ships a little more seaworthy and this would also cost the lives of thousands of sailors and that of another of the presiding officers, Rear Admiral Sir Basil Beaumont.

Yet the Admiralty must have been privately highly satisfied with Halley's performance. Less than three weeks after the court martial and following a high-level meeting with Halley, their Lordships instructed the Navy Board to have *Paramore* surveyed and to suggest improvements or if need be to allocate an alternative RN vessel for Halley's disrupted expedition.

The surveyor's report stated *'that there is noe Vessel may be more fitting that she is for that Service'* [4] and suggested some minor alterations, unaware of *Paramore's* destination. The Navy Board, in stark contrast to their sensible instructions for the first voyage, then issued orders to those responsible for provisioning and manning to fit out *Paramore* for *' a Forreigne voyage to ye West Indies & Other parts...'* This instruction was endorsed:- *'To fit ye Paramour Pinck & mann her for a voyage to ye West Indies, wth her former Complemt of men allowed her.'* [5]

Given that Halley was in reality about to embark on a voyage deep into the mighty South Atlantic Ocean (appendix 3), these instructions issued on 13th August 1699 were clearly an Admiralty smokescreen and this is confirmed by the peculiar set of events that followed.

A letter was sent from Burchett to Newton requesting an appointment for Haddock, Shovell and himself in order to discuss the longitude problem [6]. Newton replied on 22nd August [6] and they agreed to meet at the Mint on the morning of the 28th.

Meanwhile at a specially convened meeting of the RS on the 26th, Newton displayed the angle-measuring device he had loaned to Halley. This meeting was presided over by the new president, the Lord Chancellor, Lord John Somers, who had recently succeeded Montague and who was yet another political crony friend of Newton's.

The minutes of the meeting record that *'Mr Newton shewed a new instrument contrived by him for observing the moon & Starrs, the Longitude at Sea, being the old Instrument mended of some faults, with which notwithstanding Mr Hally had found the Longitude better than the Seamen by other methods.'* [7] At the same meeting Halley *'Shew'd the several Variations of the Needle he had observed on his voyage, sett out in a Sea Chart, as also he shew'd that the Coast of Brazile was ill placed in the Common Mapps, and he shew'd some Barnackles which he observed to be quick of growth.'*

These comments of Newton and Halley must rate as the most underplayed navigational statements of all time; they had both guessed the reason for the imminent meeting and hastily placed their respective claims on record in the few days left to them. Claims which they had not intended to make public until the re-started voyage had been completed, at which time a full account would have been presented to the co-sponsors as was to be the case concerning Halley's magnetic variation data. Had they intended to make a preliminary announcement they would have done so at an earlier meeting, immediately on Halley's return, as Halley had done in advising the RS of other odd biological discoveries.

There exists no record of the discussions at the meeting held two days later at the Mint although these were obviously centred on Newton's angle-measuring device and of Halley's recent and successful use of it in determining longitude on the high seas. Shovell's interest in the subject being foremost that of Halley's commander and Haddock's the information he had gleaned from attending the court martial hearing and from Burchett who had debriefed Halley.

Doubtless they questioned the Warden of the Mint as to the feasibility of large-scale production and were informed that this would be an enormously expensive undertaking; a venture that would disrupt the nation's re-coinage efforts. Limited production then, but still under the direct control of the Warden? Newton could certainly manage this; one instrument fully calibrated per month at a cost approaching £50 (£10,000) per item? Given that the navy's limited resources were already stretched, the Admiralty delegation would surely have balked at such an offer. A dozen instruments would cost nearly as much as a frigate!

Shovell, whose opinion of landlubbers and their interference in naval matters had already been made clear at the court martial would certainly have vetoed any such offer. Only a highly skilled astronomer could use such a device to determine longitude prior to the publication of a lunar almanac.

However, as all four knew, Halley had regularly used the all-weather twin-mirrored self-correcting instrument to determine *latitude* using existing solar tables and whilst they may not have been certain how accurate these readings had been, the instrument was obviously many times more accurate than a back-staff. Good enough for the poorest navigator to improve his *latitudinal* noon sight assessments up to10 fold in heavy seas, the very conditions most likely to stand a vessel into danger (chapter 23) - but far too expensive to produce.

It later transpired that they then prevailed upon Newton to keep his revolutionary invention under wraps and ordered Halley never to discuss its details; it was after all a secret weapon of immense value even if at present the Royal Navy could manage without it.

It should be noted that Halley already possessed Newton's original descriptive note and drawing (chapter 15) and also be emphasised that Secretary to the Admiralty Josiah Burchett, who was to remain in that influential post for the next 47 years, was fully aware of the efficiency of

Newton's instrument and knew that it was self-correcting and utilised twin mirrors (chapters 29 and 30).

Whether inducements were offered, who can tell? However when the current Master of the Mint, Thomas Neale died three months later, the government wasted no time in appointing Newton to the lucrative sinecure and he took office on his birthday only two days after the death of the incumbent; a position which had provided Neale with commissions to the tune of £22,000 (£4 million +) during the re-coinage in addition to his annual salary of £500. Financial compensation for accepting that his invention was placed on the secret list maybe, but for the countless poor souls who subsequently lost their lives at sea because of the delayed introduction of a reliable sea-going angle-measuring device, a tragedy.

In many ways it was also a tragedy for Newton because the sacrifice he had been forced to make would mean that he would not be able to publicise the instrument he had specifically designed to stimulate further research into the mysteries of the lunar orbit on his and his Maker's behalf.

Newton's instrument was never specifically *ever* mentioned in public again during his lifetime (but see below). Halley, about to be promoted a full (post) captain and only one rank below admiral and whom Hooke had previously considered to be spying for his country [8] during either his southern North Sea surveys or during salvage operations on the *'Guynie'* wreck (chapter 11), could clearly be relied on to keep his mouth shut. He never published any details of how he determined either latitude or longitude on any of his voyages. Just in case Halley might have thought this blackout was transient, he was later diplomatically reminded of his obligations in an official order from the Admiralty dated 23rd June 1701 [9] (chapter 21).

Halley was not to remain in England for long and he received new sailing instructions from the Admiralty (this time formulated by himself - appendix 3) and was ready to depart from Deptford again on 26th September 1699. But before he could set sail he had, once again, to sort out his crew allocation.

Of the first eight individuals to sign on, seven had been with him on the aborted voyage. One was his friend the ship's surgeon George Alfrey and another was his personal servant Richard Pinfold. However, in replacing the mutinous boatswain Dodson, Shovell and Haddock had approved the provision of one with an arm missing! At least the replacement officer, Edward St. Clair was to prove ideal (until he jumped ship) and Halley also managed to obtain an additional four crew on the strength of the boatswain's missing limb.

During those two short summer months ashore Halley had not spent much time in the company of Mary and their three children. What his unfortunate wife thought of this went unrecorded; the opinion of the spouse was rarely considered at the close of the 17th century.

He attended all possible RS meetings and apart from that historic one of 26th August, he had earlier presented the RS with a branch of a Banyan tree (29th July) and a week later it was part of a mangrove plant and then another botanical specimen. He could have presented everything at the one meeting and spent more time with his family.

Even when he was at home he spent his nights with his instruments double-checking lunar occultations. Significantly in August and September the Moon was observed eclipsing stars in the Hyades cluster, a prominent group in the Taurus constellation, on no less than three occasions; unusual events **. This allowed Halley to double-check measurements of the various stars in this cluster, which in turn enabled him to set off on his second voyage armed with a home-made Hyades almanac. He would now know that he could obtain an accurate lunar/Hyades longitudinal fix on 21st April and 9th August 1700. All he had to do was to arrange to be within a day or two of a landfall on those dates and an accurate positional fix was assured; always assuming bad weather did not interfere. This new data, when added to all the other information he had previously accumulated would surely guarantee success of his expedition this time.

Many years later in 1717 Halley *anonymously* published *'An Advertisement to Astronomers, of the Advantages That May Accrue from the Observation of the Moon's Frequent Appulses to the Hyades during the Three Next Ensuing Years'* [10] (appendix 8) which explained how longitude could be determined at sea by this method.

This voyage was not simply the old one re-started. Halley's own sailing instructions now specifically authorised him (in addition to his longitudinal and magnetic variation work) to discover land between South America and Africa and to endeavour to reach latitudes between 50° and 55°S in the Atlantic rather than sail as far south as possible in search of *'Terra Incognita'* as previously ordered. So much for the cover story *'To fit ye Paramour Pinck & mann her for a voyage to ye West Indies'.*

At some time during this busy interlude, Halley must have met the captain of *Falconberg* and agreed if possible to sail in company as far south as Madeira, the faster *Falconberg* prepared to adjust her speed to enable the Pink to maintain station. After that Halley would be on his own, the larger ship's protection no longer important; in any case *Falconberg* had a living to make despite any obligation her captain might have had in escorting a much slower vessel sponsored by the RS, an organisation which at the time held £800 (£160,000) of stock in the company which paid his wages [11].

The family were again on the dockside waving as the Pink's lines were cast off as *Falconberg* sailed slowly past. Did Mary Halley know of the terrible reputation of that vessel? Probably not. It would have been interesting to have been a fly on the wall in the building up on the hill above Greenwich to observe Flamsteed's behaviour as *Paramore* got underway for a second time a mile below.

At the very next meeting of the RS, not held until the 4th November 1699 and presided over by vice-president Sir John Hoskins, the usual chairman (Somers was to preside over only three meetings in the five years in which he held office) Hooke made a clear allegation of theft against Newton. This was noted in the minutes, which state *'Dr Hook said that the Instrument mentioned at the last meeting was of his own before ye year 1665 and that the use and fabrik of it was declared in the History of the Royal Society.'* [12] An-out-and out lie as it later transpired, but one which unwittingly provided the excuse needed to minimise the publicity Newton had deliberately attracted to his invention. In any case by now Halley and the instrument were both safely out of reach and well on the way to the island of Madeira.

To give some idea of the costs of building and operating a slave ship, in 1699 the Royal African Company paid £5 (£1,000) per ton for a 320-ton vessel the size of *Falconberg*. A ship of this displacement would be designed to carry one slave per half ton; 640 hapless souls crammed in layers into the equivalent volume of small house. They had been purchased for about £5 each and would be sold for between £12 and £23 depending on condition and demand.

Taking into account the manning and running costs, a ship would pay for itself more than twice over on one single 'middle passage' trip and make a handsome profit on the other two legs. Just so long as most of the slaves survived the crossing and the voyage was uneventful; which of course was rarely the case.

The RAC had originally attempted to be selective in their choice of Negro slaves. The company rule instructed agents to purchase only those slaves between the ages of 15 and 40 with a preference for males, but other agents were not so fussy.

When the demand for slaves increased sharply the RAC was forced to alter its buying policy, and by 1701 was permitting the purchase of boys and girls as young as 12 years of age. After all youngsters were cheaper to purchase and many more could be packed into each vessel.

However, crewing a slave ship was not without its dangers. All too often they faced serious trouble before the ship had even finished loading and mutinies and outbreaks of violence amongst the slaves were common at this point. Log book entries such as *'Condemned and burnt for killing a seaman'* highlight not only the inhuman behaviour meted out but also the desperate measures the poor wretches must have been prepared to go to because they knew that once out of sight of land their chances of ever seeing their homes again were gone forever.

Once the ship was clear of the African coast, the crew could relax a little; all the captain needed to do now was to successfully outwit pirates, avoid becoming becalmed and running out of water and hold his course to prevent running aground on some desolate coast. Should any of these unfortunate events occur, some or all of the human cargo would be thrown overboard.

Mercifully, by end of the 17th century these catastrophic events were becoming a little less frequent (discounting the piratical element) as the merchant marine navigators became more experienced in understanding some of the idiosyncrasies of the central Atlantic. Nevertheless, the logistics involved in feeding and watering a consignment of 600 for upwards of six weeks on the high seas hardly bears thinking about, and unsurprisingly the average death rate on the middle passage still often exceeded 25%. Imagine the outcry today if 25% of cattle consignments regularly died during shipment.

On the islands of Barbados and Jamaica, the RAC employed permanent agents to handle their affairs. These fearsome individuals had to be waiting on the quay when the slave ship docked because captain and crew were not above slipping slaves ashore and selling them privately. Who was to say how many had died en route? For purposes of assessing commissions, the agent was responsible for counting the survivors; any that could walk unaided were considered to be alive.

Some of the company's local representatives had acquired a reputation for extreme cruelty when dealing with difficult or un-saleable slaves, cruelty which spilled over into political life.

The agent William Beeston became the equally despotic Governor of Jamaica, a position he held for eight years until this practice of holding such dual influential offices was banned by the English parliament in 1698.

The best of these agents might expect to earn £1,000 (£200,000) or more per year from their legitimate RAC deals alone and Halley would shortly experience the high-handed behaviour of one of these men at first hand.

(1) Norman Thrower, ed., *The 3 Voyages of Edmond Halley* (London, 1981), p.286.
(2) Eugene MacPike, *Correspondence & Papers of Edmond Halley* (Oxford, 1932), p.109.
(3) E. Forbes, L. Murdin & F. Willmoth, eds., *The Correspondence of John Flamsteed* (Bristol, 1997), Vol. 2, pp.811-3.
(4) N. Thrower, ed. pp.292-3. (5) Ibid., p.293-4.
(6) J. Scott, ed., *The Correspondence of Isaac Newton* (Cambridge, 1967), Vol. 4, p.314.
(7) Isaac Newton, *Extract from Royal Society Journal Book* (16thAugust 1699).
(8) E. MacPike. p.186.
(9) N. Thrower, ed. pp. 328-9.
(10) Edmond Halley, *An Advertisement to Astronomers* (R.S.Phil.Trans., 1717), Vol. 30, pp.692-694.
(11) Henry Lyons, *The Royal Society 1660-1940* (Cambridge,1944), p.105.
(12) Robert Hooke, *Extract from Royal Society Journal Book* (25th October 1699).

CHAPTER 19

*Note. ** in chapters 17,18,19 & 21 link to Appendix 4.*

Paramore and *Falconberg* finally cleared the English southwest coast on 8th October 1699 and set course for the Madeiran island group over 1,000 miles to the south. After an uneventful passage, as soon as the two ships came within sight of the peculiar pointed mountains of Porto Santo Island, one-time home of Christopher Columbus, *Falconberg* pressed on more canvas heading directly on south towards the West African Gold Coast, soon drawing away from the smaller vessel. English slave ships did not pay social visits to Portuguese ports and *Falconberg* had time to make up.

After reaching the Cape Coast the vessel was loaded with her normal cargo of over 500 manacled wretches, but then became trapped in the central Atlantic light air belt, ran out of water and many of the slaves died of thirst. Eventually only 339 were counted off when the ship arrived in the Caribbean. The three previous cargoes had delivered 592, 491 & 502. As usual, the exact number of Negro deaths went unrecorded but this time could well have exceeded 200. Just to compound the disaster, an outbreak of fever on the island of Barbados forced *Falconberg* to sail on to the islands of Antigua and Montserrat to sell the survivors.

Halley was also to suffer loss. *Paramore* was unable to call in at Madeira owing to storm conditions and again he was denied the opportunity of fixing the island's position; then his cabin boy fell overboard! In those days ships did not carry life-saving equipment and few sailors could swim. Although an oar was thrown out immediately and the ship was brought up head to wind within minutes, the lad drowned. Manley White was the only crew member Halley ever lost and by all accounts this accident affected him badly.

Carefully skirting the Spanish-owned Canary islands again, where Halley took an accurate latitudinal fix on the westernmost island of Hierro, they called in at Sal in the Portuguese Cape Verde group. *Paramore* anchored in a bay half way down the west coast of the island and the following morning (2nd November) Halley headed a shore party to this sparsely populated part of the island, apparently to purchase salt. In truth he intended to set up his big telescope and tripod ready to take a sighting of Jupiter that evening. He knew that an hour after sunset the largest of Jupiter's moons Ganymede would suddenly become visible as it emerged from transiting the blob of a planet above the sea in the western sky **. From this single observation Halley was able to update his previous visit's DR assessment of the longitude of Sal from 22°00'W to 23°00'W. He also determined the island's latitude as being between (NE end) 16°55'N and (S end) 16°35'N. All three plots were virtually perfect and Sal was, for the first time placed correctly on the world map thanks to Ganymede (longitude) and Newton's twin-mirrored angle-measuring device (latitude) if Halley was ever going to be allowed to advertise this. The fact that he did not explain in his log exactly how he had arrived at Sal's longitude is not surprising. As captain of a Royal Naval vessel, he had landed on Portuguese territory without permission at a site not recognised as a proper anchorage.

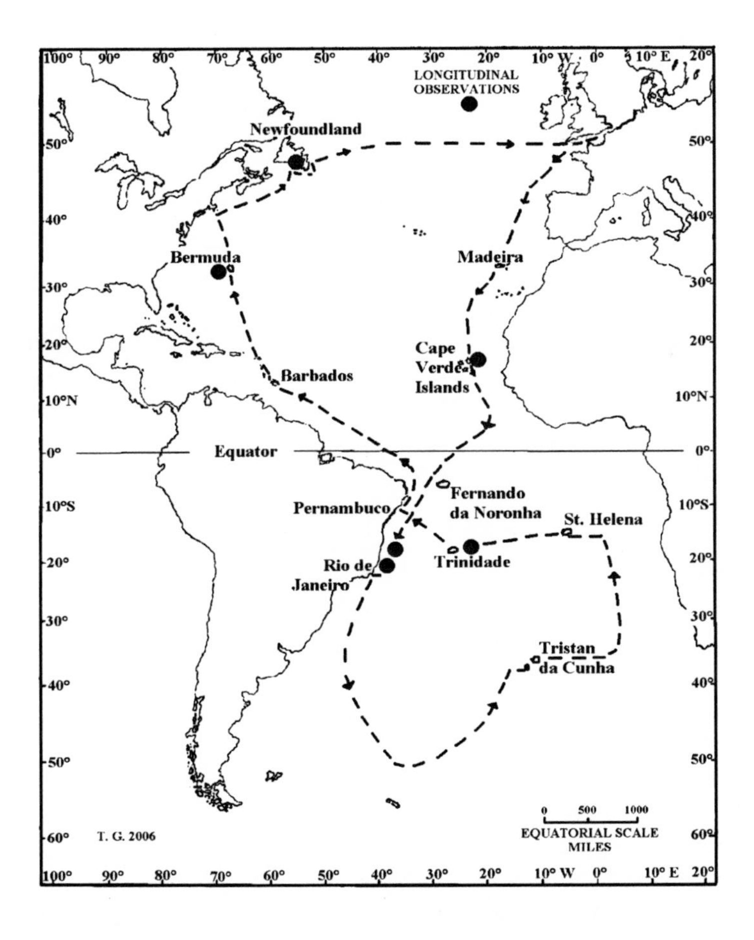

Figure 7. The track of Paramore's second voyage.

There he had set up scientific apparatus, which had extracted important navigational data that could have been put to good use by Portugal's enemies. This decision to exclude any details of his nocturnal scientific activities ashore on Sal in his log was a wise one.

The crew also took on fresh water, killed a couple of goats and caught two nesting turtles; a Hawksbill and a Green. On their previous voyage they had tried and failed to catch Green turtles on the island of Fernando de Noronha, but this time they ambushed them at night. Halley then proceeded to correctly estimate the longitude of the nearby islands of Boa Vista and S.Tiago, all based on his Sal positional plot. This enabled him to depart on his second trans-Atlantic crossing with, unlike his first, the correct departure longitude; a hasty departure nevertheless.

Having anchored 'officially' off S.Tiago, the Portuguese Governor was uncompromisingly hostile. He would only allow the taking on of water and strictly forbade any of Halley's crew to buy meat or leave the confines of the port. At least this time no one fired on them.

The first mention of a longitudinal lunar 'fix' in Halley's journal does not come until the 14th December when *Paramore* was already far south of the equator and closing in on the Brazilian coast heading for wood, water, rum and relaxation at Rio de Janeiro.

The observation of 14th December, made 35 days after leaving the Cape Verde islands, contains the most complete surviving explanation of determining longitude at sea that Halley ever provided. He mentions for the only time, that he compared the data with previous observations made in England, thus confirming his method. This is also the only other occasion he mentions time to the nearest second in any of his journals.

'This Morning [14th December 1699] *the Moon Aplyed to a Starr in Virgo of the 4th Mag. whose Longitude is* [Libra sign] *0deg 39' Lat 1 deg 25'. The Moon did exactly Touch this Starr with her Southern Limb at 3h.15' in the Morning and at 3h 20' 20" the Southern horn was just 2 Minutes past the Starr haveing carefully examin'd this observation and Compared with former observations made in England I conclude I am in a True Longitude from London at the time of this observation 36deg15' and at this noon 36deg 35'. That is according to the Accot I have of it, about 5 Degrees East of Cape Frio.'* ** [1]

Halley increased his DR longitude by a full 6° accordingly. In other words the current had pushed them west in December by roughly the same amount as it had done on the first middle passage back in March although this time *Paramore* was angling further south. He rightly deduced that this current was a permanent feature of the South Atlantic, a point that was already well understood by the Portuguese, and may have explained the hostility of the Governor of S.Tiago who had quite possibly been informed of Halley's unofficial landing on Sal.

The following night Halley was fortunate enough to obtain another longitudinal fix, this time involving the close proximity of the Moon and Mars. *'the Moon apply'd to Mars who was in a Line wth her horns at 4h.3m. or when Cor M was 8deg 6' high in the East Lat 21deg 30'.'* ** [2] Whether or not any useful data was obtained from this observation is difficult to determine as the log was not altered a second time. Either this second measurement confirmed the first, or Halley did not have sufficient information on Mars on board and hoped he might be able to check it on his return to England.

Paramore did not come within sight of the South American coast for a further five days, when Halley deduced his position, based on those two fixes plus five days of dead reckoning sailing as 22°00'S, 42°47'W. He correctly identified his position as being off Cape Thomas and his latitude was less than four miles adrift. But again he had somehow overestimated his longitude by a degree.

The ship anchored in Rio de Janeiro bay, passing the *'Sugar loaf at the entrance'* and the crew presumably took a fortnight's Christmas holiday whilst the Portuguese saw in their new century at the same time (see Notes on dates). What any of them did there is a mystery. No log entries, no mention of astro- observations, nothing.

Paramore then battled south into deteriorating weather and increasing daylight for the entire month of January without once being able to obtain any sort of longitudinal fix. By the beginning of February, Halley was commenting on the temperature in his cabin being only 11°C and a far cry from the summer temperature he was expecting. After all they were at the equivalent latitude of southern England not Greenland! He also mentioned seeing five or six different species of seabirds and began to worry that he was nearing land.

A few days later *Paramore* ran into thick fog and the prudent captain gave instructions to take depth soundings every two hours to the evident amusement of two species of penguins (probably Gentoo and King) which continually sported round the ship.

Halley carefully described them and correctly identified them as being penguins but was puzzled by their apparent lack of wings and their consequent inability to fly. Had Halley known *Paramore* was some 200 miles from the nearest land at the time, he would have been even more puzzled.

On 11th February 1700, the air temperature had dropped to 0°C and they had reached the latitude of 52°24'S. The sudden sight of three large flat-topped islands bearing dead ahead late in the evening in the midst of a full gale caused considerable concern and with difficulty the crew managed to wear ship and bear away for the night.

In the morning Halley intended to send a boat to explore; these islands were not marked on any chart and Halley was under instructions to lay claim to any new lands for the Crown. *'In the night it proved foggy, and continued so until this day at noon, when by a clear glare of Scarce 1/4 of an hour we saw the island we call beachy head* [the largest of the three islands] *very distinctly to be nothing else but one body of Ice of an incredible height, whereupon we went about Shipp and Stood to the Northward.'* [3] Halley the scientist, knew how fortunate he had been. They were not islands but enormous icebergs and had the gale forced *Paramore* much closer the previous evening, they could easily have grounded on the submerged section of one of the massive floes.

Although they had not yet reached the maximum latitude of 55°S, Halley rightly decided to get clear of these dangerous floating islands and head northeast towards South Africa on the next leg of his journey. And not before time; had he but known, they had been heading directly into one of the most treacherous stretches of water on the planet in a vessel designed for coastal waters.

As it was, their troubles were far from over. It took them nearly a week to get clear of the ice field and surely the rigorous regimen of log-line measurements and timing would have been relaxed as they dodged this way and that to avoid yet another massive block of ice that loomed up out of the mist? Certainly the entire crew must have been exhausted. The absence of a Moon forced Halley to lay to every night in these unknown seas, another very sensible precaution but one that made life desperately uncomfortable and delayed progress.

The island of Tristan da Cuhna was to be next on the agenda if Halley could find it, and if not straight to Cape Town, but the navigational problem Halley now faced was daunting. Without a recent longitudinal fix and having left Rio de Janeiro thinking he was over 1° further west than was the case, he was now aiming for a tiny group of islands that he knew were 'ill positioned' on his chart anyway. He was looking for a tight little bunch of dangerously sharp needles in a very large haystack.

Suddenly, 17 days after their near disaster with the three floating islands, at five in the morning the southernmost island of the Tristan da Cuhna group was spotted not four miles distant dead ahead. Although Halley knew the given longitude of the islands was suspect, he had still arrived in the approximate vicinity far earlier than he could possibly have anticipated. Again the currents had been at work. This time the west wind drift currents of the deep South Atlantic had pushed them as far east as the equatorial currents had earlier pushed them west.

Halley must have been very surprised to see Nightingale Island even if he did not comment on this in his log. Very surprised and very fortunate. The normal method of island finding in those days was to arrive at the correct latitude and then sail cautiously along that imaginary horizontal band until the island appeared.

However, Halley had only reached the approximately known latitude a matter of hours before Nightingale Island emerged out of the dawn right in front of him. Although *Paramore* had been under reduced canvas during the night (the Moon had set at midnight), had they arrived a few hours earlier the little ship could have been trapped on a lee shore and smashed to driftwood at the base of the island's rugged cliffs in the pitch dark. He and his crew could so easily have perished there and then and none but the teeming wildlife would ever have known their fate.

Having found the needles in the haystack even before he had started to search for them, the weather was far too stormy to risk a landing on any of the islands and they pressed on east towards Cape Town, but not before fixing the latitude of Nightingale Island to within three miles of its true position. Another remarkable feat given the sea conditions.

Then a huge storm rolled *Paramore* on her beam-ends and Halley sensibly decided to forgo Cape Town in favour of warmer and calmer waters and his next planned port of call, St. Helena. Five days after leaving Tristan an opportunity arose to observe a lunar eclipse but *'the Sky was all overcast, as is usual in these Climates'* and on the 18th March they reached the latitude of St. Helena. This time he was taking no chances, he had given himself enough sea room to run the latitude down in the time-old manner.

A further two weeks of rest and relaxation followed at St. Helena in, for Halley at least, familiar surroundings. Although it was 23 years since his previous yearlong visit, he must have met a few old friends even though he made no mention of this in his journal. This was another of those blank episodes in the voyage but the long time ashore was entirely explicable. Halley had time to spare before his planned 21st April appointment with the Hyades cluster.

Next to double check on the west-flowing current that would carry *Paramore* back to Brazil again, this time starting from the African side much further south. The reason for selecting the island of Trinidade (rather than the coast of Brazil) as the next objective was to fulfil the Admiralty's orders to take possession of undiscovered land on behalf of King William III, an assignment thwarted on the first voyage.

Not that the island was undiscovered because it was marked on Halley's chart; marked with the approximately correct latitude at least and Halley had always intended to lay claim if he could. *Paramore* headed southwest until she reached the latitude indicated, when the course was changed to west and running down the latitude began. Unfortunately they then ran into lighter winds than anticipated and their progress was slowed. So when 21st April arrived they were still well short of Trinidade.

Heading due west barely an hour after sunset and with *Paramore* at last cutting along with a steadying wind Halley managed to obtain his first longitudinal fix for 17 weeks. Next day he wrote up his log. *'Last night* [21st April] *the Moon Apply'd to the Contiguae in facie Tauri* [the Hyades cluster] *and I got a very good observation, whence I conclude my Selfe 2 degrees more to the Westward than by my Account* [i.e. 27°20'W].' ** [4]

For the next 3½ days the crew monitored their through the water speed meticulously and maintained their compass headings with equal care. This dedication to detail, so lacking in certain members of the crew of the first voyage, paid dividends. The three rocky outlying islets of Martin Vaz were spotted by the lookout dead ahead at a distance of 18 miles on the morning of the 25th which, after calculating the distance and direction run since the lunar fix, Halley was able to assign a position to - 20°25'S, 29°21'W.

The following day they closed on the island of Trinidade and cautiously dropped anchor a mile from the shore on the lee side. Halley then estimated Trinidade's position as being 20°25'-29'S, 29°50'W. Remarkably in both instances his latitudes were spot on and he was only 30' in error in both longitudes. Positions far more accurate than for any remote island at that time. Not as accurate as his Sal plot where he had made use of his big land-based telescope but more accurate than his positions for either the coast of Brazil or the island of St. Helena. Halley's private Hyades almanac had proved its worth, not of course that he could be certain.

With considerable difficulty owing to the high winds and rocky shoreline, they landed on 4- mile-long Trinidade, hoisted the Union flag and took possession in the name of King William III. Halley left *'some goats and hoggs and a pair of Guiney hens'* brought from St. Helena in the hope they would breed and leave a fresh meat supply for future mariners. They managed to replenish their water casks, *'emptying my Cisterns of their brackish St Helena Water wch by reason of the great Rains whilst I was there, was so turn'd sometimes as not to be fitt to take on Board'* [5] before upping anchor and setting course for the South American mainland.

A further eight days of sailing NNW carried *Paramore* to the Brazilian coast and the port of Pernambuco (Recife) exactly as planned. Here was another opportunity to survey a major Brazilian port but Halley's luck was out again. He was prevented from obtaining any longitudinal measurements through lack of opportunity and prevented from surveying the port, not by the Portuguese this time, but by, of all people, the local Royal African Company representative!

Having been regally entertained by the local Governor but nevertheless forced to anchor alongside two overpowering Portuguese battleships, Halley was introduced to the English Consul. This gentleman had Halley arrested, held on shore under armed guard and then searched his ship despite being supplied with papers showing Halley held the King's Commission and knowing the ship was a Royal Naval vessel (appendix 5).

The Governor sensibly intervened and it then transpired that the 'Consul' was nothing of the sort; merely another of those RAC men wielding frightening power, this time in a foreign country! Here was a private English citizen claiming to be a Government official actually arresting a Royal Naval captain and searching a Royal Navy vessel in a foreign port. One can only presume this man was in Pernambuco to supervise the unloading of slaves from RAC vessels; if so a highly secretive operation.

Had Hardwick, for that was the man's name, whilst rifling through the ships papers hoping to find evidence of illicit slave trading, discovered that Halley had illegally established the longitude of Portuguese islands in the Cape Verde group, a diplomatic incident could easily have resulted. A far cry from merely landing on the uninhabited island of Trinidade and taking possession in the name of his king. Halley's fixing the position of islands that were regularly used as a starting point for the trans-Atlantic crossing to Brazil could make the crossing safer for Portugal's enemies. Whether or not Hardwick examined Newton's angle-measuring device, which must have been on board at the time, is not known. Halley left for the Caribbean.

The north-flowing current aided their passage and they docked in Bridgetown, Barbados, the most easterly of the Caribbean islands on 1st June 1700 almost exactly 10 months after the visit by *Paramore* on her ill-fated first voyage. At least this time there had been no mutinous crew trying to prevent the stopover. But once again all did not go smoothly; this time Halley caught a fever; *'I found my Selfe Seized wth the Barbadoes desease..'* [6]

Edward St. Clair conned the ship out of the Bridgetown anchorage at the earliest opportunity and passed close by Antigua where *Falconberg* had probably recently unloaded most of her human cargo. *Paramore* dropped anchor at St. Kitts but because Halley was so ill, several opportunities to observe Jupiter were missed. Whilst either Alfrey or St. Clair could probably have handled the observations, everyone's minds were rightly focused elsewhere.

Halley was still desperately ill when *Paramore* left the Caribbean heading for Bermuda, that tiny dot of an island far off the coast of North America in the middle of the NE flowing mighty Gulf Stream. At least this was one current that had been reasonably well understood ever since Columbus first unwittingly used it to help him get his weed-encrusted worm-riddled ships back to Europe. The island of 'Bermoodas' was reached in Halley's usual no nonsense manner, the low-lying southern tip first being spotted to the NW 15 miles distant. This would indicate his navigation was again very accurate but again a mite too close for comfort. No running down the latitude this time either.

During his stay on the island, Halley made a number of observations, but infuriatingly he simply states *'I made Several observations to determine the Longitude of this Island, which I find to agree with my reckoning account* [his DR from Barbados], *vizt that the Longitud of St Georges Island is 63deg 45' ... and the latitude thereof 32deg 24.'*.[7] At least he had recovered sufficiently to organise the setting up of his long telescope and make observations; on many nights Jupiter was ideally placed. As usual, his latitude was accurate but somehow his longitudinal error had crept back to a full degree, which is odd because he used a fixed telescope and the Jovian moon method, although it must be remembered he was still an extremely ill man and he had just lost St. Clair who had decided to jump ship at this point.

It was from here in Bermuda that Halley despatched his letter to Burchett mentioning he had been stricken with the Barbados disease which also stated that *'to morrow I goe from hence to the coast alongst the North America and hope to waite on their Lordsps: my selfe within a month after the arrivall of this, being in great hopes, that the account I bring them of the variations and other matters may appear soe much for the public benefit as to give their Lordsps intire satisfaction.'* (appendix 5)

The *'other matters'*, the claiming of Trinidade for the Crown and the numerous longitudinal plots, were not specifically mentioned; possibly because the letter could have fallen into enemy hands en route. Significantly Halley was also careful to notify their Lordships at the

Admiralty of his brush with Hardwick at Pernambuco in the same letter:- *'Mr. Hardwyck that calls himselfe English consull, shewed himselfe very desirous to make prize of me, as a pyrate and kept me under a guard in his house, whilst he went aboard to examine, notwithstanding I shewd him both my commisions and the smallness of my force for such a purpose.'* A perfectly understandable notification, but if Hardwick had examined Newton's secret instrument, also a sensible precaution.

After careening and re-caulking *Paramore* and another very wise spell of rest and relaxation ashore, Halley set course for Nantucket, planning to sail on up the North Atlantic seaboard towards Newfoundland; an important landfall whose longitude was suspect. He had another appointment with the Hyades cluster and needed to get *Paramore* close to a coastline by 9th August.

The anticipated night arrived one day before sounding shallow water just south of Newfoundland; perfect and Halley's luck was in, the night was clear. *'This morning I observed the Moon aply to ye Hiades.'* ** [8] and he adjusted his DR longitude slightly.

Halley was now in stormy waters often enveloped in fog where he came across a fleet of English fishing boats that took *Paramore* for a pirate and sensibly fled. On trying to anchor in a small harbour for the night *'one Humphrey Bryant a Biddiford man fired 4 or five Shott through our rigging but without hurting us.'* [9] Having given the Devonshire fisherman a severe reprimand Halley decided to forego any further survey work in the area and set course for the Scilly Isles.

Still 600 miles short of landfall, the lookout reported a sudden change in the sea colour, a sure sign of shoaling water. But the deep-sea lead found no bottom at 75 fathoms and Halley concluded they were passing the edge of an uncharted bank. Modern contour maps show no such bank and almost certainly *Paramore* had sailed through an area of convergence where the Gulf Stream meets the colder north Atlantic waters. Halley gave the ship's position as 50°08'N., 23°30'W., the first ever assessment of the northern limits of the Gulf Stream.

The trip across to the Scilly Isles landfall was otherwise uneventful but no longitudinal fixes were possible during the entire 19-day passage. Passing to the south of the Scilly Isles Halley also obtained a first-class latitudinal observation which correctly placed the southernmost dangerous reefs at 49° 50'N *'past Dispute.'* [10]

On the 12th September 1700 a somewhat battered little ship anchored in the Downs but Shovell was away on royal escort duty so the Pink proceeded to Deptford where *Paramore* was laid up and the crew paid off.

Also in that letter from Barbados to Burchett were details of their lucky escape from the jaws of death that had *'cost me my skin'* yet the only mildly appreciative note received from the Admiralty immediately following the return of a highly successful scientific voyage was in a letter from Burchett which contained the paying off instructions. *'... I dare assure you that you will not offend in comeing to Towne;* [before being given official permission to leave his ship] *only lett me give you this Caution, To have ye Books in readiness and to attend to the payment of Vessell.'*

Burchett also at least had the good grace to apologise for the poor upwind sailing abilities of *Paramore* which were supposed to have been remedied prior to departure[11].

Although Halley handed his accounts and log books over to the Admiralty and obviously was thoroughly de-briefed, nowhere in Admiralty records is there so much of a hint of any discussion regarding his new and highly successful methods of determining either latitude or longitude whilst actually on the high seas.

Halley's verifiable positional assessments from both Atlantic voyages are set out in detail in appendix 4. These include 11 longitudinal plots, none more than 1° in error and 14 latitudinal plots, only one of which was more than 0° 05' (five miles maximum) in error.

(1) Norman Thrower, ed., *The 3 Voyages of Edmond Halley* (London, 1981), p.144.
(2) Ibid., p.145.(3) Ibid., p.163.(4) Ibid., pp.182-3.(5) Ibid., p.185 (6) Ibid., p.307. (7) Ibid., p.199. (8) Ibid., p.204. (9) Ibid., p.205. (10) Ibid., p.210. (11) Ibid., p.314.

CHAPTER 20

Despite Edward Harrison's original scant report [1] that had included some information on his magnetic variation observations but sensibly contained not a word about Halley's successes, it must have been painfully obvious to Flamsteed that he had lost out. Employed by successive monarchs to *'with the most exact care and Diligence to rectifying the Tables of the Motions of the Heavens, and the places of the fixed Stars, so as to find out the so-much desired Longitude at Sea, for perfecting the art of Navigation'* [2], he had been bested by his two least favourite persons, in part because he had failed to stick to his appointed task.

Halley, with Newton's help, had, it seemed, beaten him in the race to establish longitude on the high seas as the pair had announced at that between-voyage Royal Society meeting, although he had no way of confirming the truth of this claim. Or infuriatingly exactly what method had been used; everyone involved had suddenly become strangely reticent.

At about the time that Halley and his crew were enjoying a few days rest and tropical downpours on the island of St. Helena, Harrison again contacted Flamsteed. This time the Astronomer Royal was furnished with Harrison's private journal; his biased account of *Paramore's* thwarted expedition. Still sensibly no mention of Halley's successful use of Newton's instrument though; Harrison was well aware of the length of the Admiralty's reach. Flamsteed carefully copied out a full list of dates, latitudes, longitudes and all the notes on magnetic observations [3]. Notably this second list of magnetic variation data contained twice as many as the first, filling in a number of important gaps.

Jealously aware that the RS were going to publish Halley's magnetic variation chart which he assumed would contain data collected by Harrison and others as well as by Halley himself, Flamsteed now began referring to his hated rival as *'Captain Raymer'*, an oblique but libellous comparison to Nicholas Reimers, the one-time Imperial Mathematician to the Holy Roman Empire, the man Tycho Brahe successfully prosecuted for theft of data and in doing so hounded to a premature death.

'A New and Correct CHART Shewing the VARIATIONS of the COMPASS in the WESTERN AND SOUTHERN OCEANS as observed in ye YEAR 1700 by his Maties Command' by Edm. Halley' [4] was published early in 1701 and was later updated several times. This owed nothing to Harrison who unbeknown to Flamsteed, had copied some of his data from Halley's

own notes. However, the chart's title did suggest it had been compiled entirely from Halley's observations. This was not the case as anyone examining the chart could see because *Paramore's* track was superimposed and went nowhere near, for example the west coast of southern Africa. In these cases Halley had obviously drawn a series of isogons based partly on other's data he had previously published [5] and partly on a projected 'guesstimate'.

Significantly, this publication carried official Admiralty approval and coastlines and islands were quietly repositioned to tally with Halley's new data, but for some reason (another Admiralty blackout?) Halley never published the vitally important results of his assessments of Atlantic surface current speeds and directions. Although the Scilly Isles were repositioned just below the prominent 50° line of latitude on the chart, one would have needed a strong magnifying glass to confirm this.

(1) E. Forbes, L. Murdin & F. Willmoth, eds., *The Correspondence of John Flamsteed* (Bristol, 1997), Vol. 2, pp.811-813.
(2) Francis Baily, *An Account of the Revd. John Flamsteed* (London, 1835/1966), pp.111-2.
(3) E. Forbes, L. Murdin & F. Willmoth, eds. Vol. 2, pp.783-4, 811-813.
(4) Norman Thrower in *The Quest for Longitude* (Harvard, 1996), Figure 9, p.58.
(5) Edmond Halley, *A Theory of the Variation of the Magnetical Compass* (R.S. Phil.Trans., 1683) Vol. 13, pp.208-21.

CHAPTER 21

*Note. ** in chapters 17,18,19 & 21 link to Appendix 4.*

At about the time that Halley's Atlantic chart was published, an anonymous two page note appeared in *Philosophical Transactions*; *'An Advertisement Necessary for All Navigators Bound up the Channel of England'* [1] (appendix 6), which was based on a statement made by Halley to a Royal Society meeting on 16th February 1701. Halley's reluctance to be formally named stemmed from the fact that he was still a commissioned officer in the Royal Navy. Serving officers might be allowed to publish isogonic charts with Admiralty backing using their own names, but they did not publish public warnings to mariners, especially not those within the Shovell orbit of influence.

The RS publication pointed out that the latitudes of the Scilly Isles and Lizard Point were incorrectly noted in pilot books, both placed some 15 miles north of their true positions.

The northerly 'indraught' up into the St. Georges Channel which was often blamed (as Narbrough had done in 1673 - chapter 10) for running into the Scillies was shown to be imaginary except in special climatic conditions. The true explanation was the incorrect positioning in pilot books of those very hard lumps of rock. The anonymous author also pointed out that the magnetic variation had *'become considerably Westerly, (as it has been ever since the year 1657) and is at present about 7½ Degrees.'*

Having issued his well-meant warning, Halley commenced a third voyage in June 1701 as commander of *Paramore,* armed as before with her six small cannons and two swivel guns. Halley would have commenced this voyage in May but for the usual manning problems. However, the Admiralty considered this expedition so important that Halley was actually given permission to recruit from the main battle fleet.

This time he was not being asked to achieve near miracles but to survey both sides of the English Channel.

He was to pay special attention to tidal races, the charting of rocks and to note magnetic variation along the treacherous Brittany coastline; all this whilst France and England enjoyed a brief period of peace before the inevitable war broke out again. In truth this was a covert mission and his orders obliquely reminded him of his obligations regarding the publishing of any information which might assist his country's enemies:- *'And in case dureing your being employed on this Service, any other Matters may Occur unto you the observing and Publishing whereof may tend towards the Security of the Navigation of the Subjects of his Majtie or other Princes* [allies] *tradeing into the Channell you are to be very carefull in the takeing notice thereof.'* [2]

Surprisingly the English south coast was yet to be surveyed, either on shore or from the sea, and the pilot books mariners relied on were desperately inaccurate.

There was a growing suspicion that the entire south coast of England was somehow artificially stretched as another paragraph in his Admiralty orders underlined: *'And you are to take the true bearings of the Principal head Lands on the English Coast one from another and to continue the Meridian as often as conveniently you can from Side to Side of the Channell, in order to lay down both Coasts truly against one another.'* [2]

An impossible assignment in the absence of a very efficient angle-measuring device of course. On the other hand, the opposing northern border of France was known to have been accurately surveyed on shore by the Jovian moon's method back in 1679 by Picard and de La Hire (chapter 9). As distances between France and England were short, these discrepancies caused nearly as many problems to navigators as did the little understood magnetic compass variation.

Although Halley's task was formidable, he completed it with a minimum of fuss and yet again his journal [3] only told part of the story. A wealth of tidal flow data, 15 painstaking magnetic variation assessments and 15 noon Sun sights to determine his latitude, 12 of which can be verified ** . None of these was more than 0°04' (four miles) in error, which confirmed beyond any doubt that Newton's twin-mirrored instrument was once again in Halley's possession. But yet again not so much as single mention of this in Halley's journal (which only came to light in 1877). He completed his English Channel survey by October when *Paramore* was paid off and the pair parted company for the last time.

Immediately upon Halley's return to London in mid-October he rendered Josiah Burchett an account of his summer's expedition and despite the Admiralty secrecy order, then took it upon himself to issue a second anonymous warning to mariners. The first section of this second warning was word for word identical to the RS publication but then went on to provide amended compass directions for westbound ships as well. Halley ended his advisory notice with these prophetic words; *'If this Notice be thought needless by those, whose Knowledge and Experience makes them want no Assistance; yet if it may contribute to the saving of any one Ship, the Author thereof is more than recompensed for the little pains he has taken to communicate it.'*

This *'Advertisement Necessary to be Observed in the Navigation Up and Down the Channel of England communicated by a Fellow of the Royal Society.'* was typeset by the printers of his first article in larger type in poster form, known, appropriately in this instance, as a 'broadside' [4] (appendix 6). The sheet was distributed to coffee houses and other establishments in London frequented by marine navigators where it was put on sale price two pence or one half of the price of a dish of coffee. Now there were two anonymous Halley publications that contained information that was of vital importance to the navigators of ships entering the western approaches to the English Channel; the narrowing bottleneck having the deadly Scilly Isles and the dangerous Lizard Point along its northern boundary. In view of the impending publication of Halley's Admiralty-approved Channel survey results by the London cartographers Mount and Page, the issuing of another anonymous warning might have been deemed pointless.

However when the two-part chart titled *'New and Correct Chart of the Channel between England and France'* was published early in 1702, the Scilly Isles and Lizard Point were still in the wrong places and most of Halley's new latitudinal positions were not properly registered - Plymouth was placed 10 miles north of the port's true position for example. Even the well-established position of Greenwich was incorrect. There was no specific mention of the 15 magnetic variation plots taken although the variation in the Channel was noted as being 7½° W. at three widely spaced locations. Remarks on the easterly chart almost certainly led to incorrect conclusions relating to inshore tidal assessment by anyone using it without additional coaching. To the confusion of England's enemies (and everyone else), the latitudinal scale was incorrectly spaced.

Strangely, unlike his two Atlantic voyage journals, Halley made not one single reference to longitudes in his private Channel journal; no dead reckoning estimates even. Although nights were short during the early part of his voyage, *Paramore* was very conveniently off the South Foreland, Ushant and at Spithead on the four nights when Halley could have made use of the Hyades cluster. On the six occasions when he could have registered lunar occultations or appulses, *Paramore* was at anchor at Plymouth, St. Helens Roads and Portsmouth; all strategic locations with suspect longitudes. On two nights when he could have made shore-based observations of Jovian moons events, he was thwarted by heavy weather and strong winds and expressed his annoyance at the conditions without

mentioning why.
Halley was now in home waters where parallax (Glossary) differences were minimal and even more importantly verifying observations could be taken every night either in Oxford or London on his behalf. The longitudinal assessments he surely made must have been accurate to within 15' *after* comparing notes against the master data. Which might explain the absence this time of on-the-spot longitudinal statements in Halley's working journal.

Halley's *'New and Correct Chart of the Channel between England and France'* does not even carry a longitudinal scale. The Admiralty had clearly placed an embargo on the full publication of Halley's results and, knowing this he had seen fit to issue his second anonymous warning. Whether the Admiralty made a point of notifying the fleet navigators of the true results of Halley's spying activities is not known, but subsequent events would suggest otherwise.

Whilst the crew of *Paramore* were being paid off at Deptford, the crew of Dampier's ill-fated Pacific expedition had coincidentally disembarked at the same place but without their ship! Dampier too had been given a rough ride by his First Officer but rather than play it by the book as Halley had done, he had unceremoniously dumped George Fisher in Brazil and continued without him.

The death of a crew member whilst locked in the ship's brig, scurvy brought about by a lack of fresh fruit and vegetables, and general lack of respect for a harsh taskmaster had all caused serious problems. However, the final straw came near the island of Ascension on the way home when *Roebuck* sprang a leak and sank; Dampier losing many books, papers, and some of his precious botanical specimens.

The crew spent six weeks on the island before being picked up by a squadron of passing English ships. It was not until June of 1702 that the ex-pirate faced a court martial at Spithead presided over by Admiral of the Fleet Sir George Rooke who had by now presumably repaid Shovell the money he had borrowed 20 years earlier.

Rooke with Shovell sitting beside him at the long table in the great cabin of *Royal Sovereign* acquitted Dampier of charges relating to the loss of *Roebuck* and the death of the seaman but found him guilty of 'very hard and cruel usages towards Lieutenant Fisher'.

He was fined all his back pay and adjudged no longer fit to be employed as a commander of any of Her Majesty's ships. A second grossly unfair judgement involving Shovell and one that reinforced his biased opinion of amateurs and ordinary seamen commanding RN vessels.

On no less than four separate occasions Halley's commander-in-chief Shovell had now been given the opportunity to take note of the correct latitude of the Scilly Isles. Yet pilot books were still being supplied to the Royal Navy in 1707 that placed the Scilly Isles and the equally dangerous Lizard Point at the wrong latitude.

But at least his sovereigns had appreciated Halley's efforts. William III ordered that he receive, in addition to his pay, a special award of £200 (£40,000), but unfortunately fell off his horse and died. His sister-in-law Anne became Queen, paid Halley the money and promptly sent him overland to the Adriatic on yet another delicate diplomatic mission.

The Adriatic ports of Trieste and Buccari were being eyed by the English as possible Mediterranean bases for their fleet. The only problem was that the Royal Navy possessed no charts. Halley was called on to conduct secret surveys, but this time it was to be done using small boats and organised from ashore.

By all accounts Halley handled this with his usual tact and efficiency and the charts were handed over to the Admiralty for the use of the commanders of the Mediterranean fleet, one of whom was Shovell. As far as is known, if issued they were never made use of; Halley's expenses were on this occasion met from Government secret service funds.

Falconberg was to complete two more slave transporting voyages. During voyages in 1701, which had commenced shortly before *Paramore* began her refit for the Channel survey, and her last in 1703 she continued to make a profit for the Royal African Company. Another 834 slaves were delivered to Barbados and Jamaica on these two runs and the travel-weary vessel tied up in England for the last time in the summer of 1704 having sailed in all some 100,000 nautical miles.

Her ninth and final voyage was to the scrap-yard, and the hulk of *Falconberg*, the most successful of all the slave ships plying the infamous middle passage, was *'cast away near Bristol'.* [5] During the last seven of her eight triangular Atlantic voyages which were completed in a record-breaking 10 years, *Falconberg* had unloaded 3,249 live slaves at Caribbean ports. What terrible memories must have become locked within her hull and one can but wonder the eventual fate of her captain. Did he sleep easily in his hammock, or did he wake screaming and soaked in sweat in the dead of night for years to come?

Little *Paramore*, only ¼ the displacement of *Falconberg* fared slightly better. After her surveying duties under Halley's command, she was refitted, re-armed, provided with a 'proper' commander and sent off to join the real Royal Navy to act as a bomb ketch; a glorified floating multiple rocket launcher. Almost unbelievably she was assigned to the Mediterranean fleet where she again came under the direct orders of Shovell. After the Navy had finished with her, she was sold for £122 (£240,000) in 1706 at a 'pin in a candle' auction to a Captain John Constable, a merchant navy officer. After that? Who knows?

Nowadays the pair of vessels would have been preserved for posterity at the National Maritime Museum just down the hill from the Royal Greenwich Observatory; *Paramore* representing good, *Falconberg* evil and both having direct links with superb oceanic navigators.

(1) Edmond Halley (anon.), *An Advertisement Necessary for All Navigators Bound up the Channel of England* (R.S. Phil.Trans., 1701), Vol. 22, pp.725-6.
(2) Norman Thrower, ed., *The 3 Voyages of Edmond Halley* (London, 1981), pp.328-9.
(3) Ibid., pp.220-247.
(4) Colin Ronan, *Edmond Halley; Genius in Eclipse* (London, 1970)
(5) K. Davies, *The Royal African Company* (London, 1957), p.189.

CHAPTER 22

During the 17th century, many of the leading European slave-trading nations maintained holding forts on the West African coast where the Negroes they had purchased from Arab and African traders could be safely held to await the arrival of a slave ship.

Denmark experienced particular problems with its holding forts, and is of special interest because of that country's direct links with English royalty and the Royal African Company. In 1658, Danish warships attacked and occupied a Swedish fort on the African Gold Coast. Hardly had they started shipping slaves from it when the fort was captured by the English. So the Danes built one of their own; Christiansborg, named in honour of their king.

Denmark had originally sold slaves to third parties, not having territory in the Americas but in 1672 they took over the West Indian island of St. Thomas and Danish involvement in the trans-Atlantic slave trade swung into serious action. They then built another fort, Frederixborg, and annexed the West Indian island of St. John in 1683.

1683 was also the year in which the Danish Prince George, the youngest and somewhat simple son of their monarch married Anne, the apparently simple but kindly daughter of the then Duke of York, who at that time was the Governor of the RAC and was soon to become James II. Having married the future queen of Great Britain, George seems to have done little for two decades other than to successfully impregnate his unfortunate wife no less than 18 times; sadly only three of their children lived for more than 24 hours.

Late in the year of the royal marriage, the RAC claimed that the Danish Gold Coast fort of Frederixborg had been built illegally on their land; a vast tract originally purchased from the local king for a few pots and pans and strings of beads by English traders way back in 1650. The RAC, after considerable diplomatic pressure on the Danes eventually took possession and re-named the holding station 'Fort Royal'. Akin to rubbing salt into a wound now that the 'royal' in question was also Governor of the company which had annexed the property. By now the RAC was top-heavy with royalty; James was king and the next in line were his daughters Mary and Anne, who were also both stockholders.

The other Danish fort, Christiansborg, was sensibly mortgaged to the

RAC at the same time, but was redeemed four years later after James' exile when the Protestant and sympathetic William and Mary were enthroned. But before continuing the story it is only fair to point out that at the end of the 18th century Denmark was the first established sovereign state to abolish slavery and one of the reasons that triggered this decision were the horrifying published figures for deaths of slaves shipped out of their Christiansborg fort during the middle passage to the West Indies.

Because Shovell had, over the years, established something of a reputation for escorting royalty, he was chosen to accompany Prince George to the Mediterranean in 1691. Officially the Prince was going to lead a combined English and Dutch fleet with Shovell as his second-in-command. The plan was to relieve pressure on the allied-held seaport of Nice, which the French were trying to recapture. The Admiralty sensibly abandoned the idea, probably fearing for the life of the Prince who had absolutely no experience of naval warfare.

The asthmatic Prince George was left waiting quietly in the wings, watching his wife's physical condition deteriorate. When their son Prince William of Gloucester, by then third in line to the English throne died at the age of 10 in 1700, George feared for her sanity. By the time of her coronation on St. George's Day in 1702 Anne was 37 years old and her consort 11 years her senior.

The new queen was too lame with rheumatism to walk any distance and had to be carried right to the door of Westminster Abbey. Somehow, she managed to walk to the throne and the dogged determination she displayed that day surprised all those politicians who had assumed she would be easy to manipulate. During her short reign, Anne was destined to become a decision-maker par excellence, and Winston Churchill was later to offer his opinion that she was one of the toughest personalities ever to occupy the English throne.

Not all her decisions were governed by common sense however; appointing her beloved George to the lofty post of Lord High Admiral of the mighty English Royal Navy was an unmitigated disaster. The poor man proved to be not up to the job and Queen Victoria was later to describe him as 'the very stupid and insignificant husband of Queen Anne.'

England was once again at war, this time allied with the Dutch against her usual enemy, France and now Spain again and naturally Shovell was deeply involved; his new queen and her consort were determined to use the might of the Royal Navy to improve England's finances.

Having dealt with Dampier's court martial, Admiral Sir George Rooke sailed from Spithead in command of a large fleet of warships and army transports. The intention was to capture the vital Spanish port of Cadiz with 14,000 troops and create mayhem all along the Spanish Atlantic coastline. If Cadiz proved impregnable they were to attack Vigo, Corunna or even Gibraltar.

Shovell was to remain in the western approaches to the English Channel to protect Rooke's back should the French Atlantic fleet chose to sail from Brest. Hamstrung by lack of ships and the men to man them, Shovell could do little all summer but organise raids on odd enemy vessels and complain to the Admiralty about his genuine manning problems. Nevertheless, his squadron did somehow manage to capture a number of unfortunates, including one with a prize value approaching £60,000 (£12 million).

Rooke spent most of the summer of 1702 in his cot suffering from gout, trying to capture Cadiz with forces more at odds with each other than their enemy. Meanwhile, information had reached the Admiralty that a Spanish Plate fleet which had left the West Indies bound for Cadiz and now had a French escort, had altered course and was now heading for either Corunna or Vigo. Rooke decided Cadiz was a tougher nut to crack than had been supposed and sailed for Vigo 350 miles to the north.

Shovell, galvanised into action by the news, now found the manning strength to sail south, hoping to intercept the Plate fleet before it reached a Spanish port. The Spanish and French neatly sailed between the two English fleets without being sighted by scouting frigates from either, docked in Vigo and began unloading the treasure and carting it into the hinterland in frantic haste.

Rook's fleet arrived before all the treasure had been unloaded, destroyed many of the Spanish ships and captured between £1,000,000 and £1,750,000 (£200 to £350 million) in silver bullion, together with a small quantity of gold. The Admiralty issued strict orders for the protection of this treasure and of any seaworthy captured vessels, and Shovell was made responsible for bringing all the loot back to England post haste, which he did after removing the armaments (including 60 bronze cannons) from any un-seaworthy enemy vessels before setting them on fire.

Still being technically in charge of the coins and bullion, Shovell personally supervised its safe delivery to the Mint at the Tower in London, handing it over to the Master, Isaac Newton. Much of the silver was distributed as prize money, but the Mint finished up with some £13,000 worth of silver pieces of eight and several hundred pounds weight of the gold.

Newton, aware of the victory at Vigo, had already altered the dies for the proposed first issue of Queen Anne silver and gold coinage to include the word "VIGO" on the obverse side under the queen's bust. A nice touch and a thumb to the nose at the French and Spanish. The day following the delivery, Newton set his foundry workers to melting down the pieces of eight and the gold bars in preparation for refining the melts, casting and rolling strips and die stamping coins of seven denominations; Crown, Half-Crown, Shilling and Sixpence in silver and Five Guinea, Guinea and Half-Guinea in gold.

Shovell then petitioned Queen Anne for a share of the bounty, citing the skill involved in bringing home the treasure safely in adverse circumstances. Officially this was rebuffed but unofficially (so as not to encourage similar claims from Shovell's subordinates) he was rewarded. For one thing, he was allowed to retain three wonderfully ornate bronze cannon taken from the French flagship, but only after the Admiralty had impressed their anchor mark of ownership.

Treasury proposals (Newton apparently believed these had been suggested by Shovell) to use the bronze from other captured cannon to produce a million VIGO halfpennies for publicity purposes was successfully repulsed. Newton explained that this task would occupy the Mint's entire workforce for a year and delay the planned introduction of Queen Anne coinage. Far better that they should concentrate on producing large numbers of the popular shilling coins (one twentieth of a £) these being in very short supply. Which is why some of the first coins of Queen Anne's reign to be issued were the 1702 Vigo silver shillings, coins that now fetch upwards of 15,000 times their face value.

Because a portion of Newton's salary was based on the value of coin minted, he had stood to lose financially had he been forced to accept Shovell's proposal; one interference from that quarter had been more than sufficient.

All the Vigo coins had grained or embossed lettered edges produced by a special secret process to prevent 'clipping' of coin. Although this process was in use before Newton became involved with the Mint, it was during his tenure, first as Warden and then as Master that the coinage of England and Scotland was completely reorganised. For his inspired work, Newton was knighted in 1705 by Queen Anne. He had also been handsomely compensated, unlike his assistant Edmond Halley, who had come close to resignation whilst trying to control corruption at the temporary Mint at Chester back in 1696. Master of the Mint Newton's gross salary for the single year 1701 was £3,500 (£700,000) although this was exceptional. For the 27 years he held the post (until his death) his average annual salary plus commissions amounted to almost £1,000.

Evidentially everyone involved in the edging process (including Halley) was obliged to swear an oath of secrecy, a secret so well kept that now no one knows exactly how the machinery operated. A second significant point is that the graining dies were engraved to an incredibly high degree of accuracy; certainly good enough for the purposes of engraving scales on high quality marine angle-measuring devices, as is made clear by the perfection of the angled graining lines on the small sixpenny pieces (chapter 30).

Having successfully destroyed or captured a considerable portion of the Spanish and French fleets at Vigo in 1702, the next year found the English Royal Navy in control of the high seas but the usual trouble in getting the sailing season under way and organising rendezvous, meant that the large and powerful fleet, now under the direct command of Shovell, left England for the Mediterranean more than a month later than planned. This created the now all too familiar domino effect, which placed the Anglo-Dutch fleet a very long way from home and still off the coast of Italy in October with their season's programme uncompleted.

Shovell was personally responsible for much of this delay. On reaching the Italian port of Leghorn (Livorno) in late September to take on food and water for the homeward journey he took umbrage at receiving only a 5-gun salute of welcome from the Governor and spent several days posturing and threatening the port with his massive firepower. The city's forts eventually fired an 11-gun salute to the Queen's flag and no less than 23 more for Shovell.

The victualling then took an age, the local population probably being none too co-operative by this time, and the fleet departed three days later than the pre-agreed last possible safe date. Under orders, the Dutch contingent promptly hightailed it for home but Shovell then became involved in further gunboat diplomacy and flag waving en-route west along the North African coast.

On 5th November the various squadrons of Shovell's fleet were all still in the vicinity of Gibraltar by which time everyone, notably Shovell, had previously agreed they should have been back home. On the 27th November, having on this occasion avoided the Scilly Isles, Shovell sailed on past Spithead and anchored off the Downs in the English Channel, an insecure but convenient assembly point. There he awaited the arrival of his other squadrons before taking pilots on board and preparing to proceed to winter quarters in the Medway and Thames rivers.

A full week later eight ships of the fleet with Shovell in the van, set sail again but only managed to round the North Foreland before running into head winds, which forced these cumbersome leviathans to anchor in the open reaches of the Thames estuary. The south-westerly winds increased to gale force and the gale force winds increased to storm force. Still the wind strength mounted, the bottom dropped out of the barometer and the greatest storm in English recorded history hit the southern counties. Winds exceeding 160 kilometres an hour smashed across the exposed Thames anchorage, ripping off canvas and deck housing, breaking anchor flukes and forcing captains to hack down massive masts and rigging to reduce top hamper and avoid capsize. Huge men-of-war were blown like so much debris straight out across the sand bars of the Thames estuary.

Miraculously all eight ships scraped and bumped their way across these sandbars without being wrecked and were blown on out into the southern North Sea, where huge waves were whipped up to more than 20 metres in height. Shovell, in the 100-gun first-rate *Triumph* had by now lost his anchors, tiller, rudder and his captain had been forced to order the main mast cut down. As the storm eventually blew itself out *Triumph* managed to rig jury masts and rudder and limp back into the shelter of the Thames. Shovell was one of the luckier ones. *Association*, another first-rater with over 600 terrified crew on board, was blown clear across the North Sea and ended up on a Swedish beach.

Those of the fleet that had remained at anchor in the Downs fared far worse. No less than 12 men-of-war dragged their anchors or had their thick holding cables suddenly snap like stems of straw. All 12 were blown onto the Goodwin, Brake and Burnt Head sandbanks where four of them broke up, exactly in the manner as predicted by Halley.

Rear Admiral Basil Beaumont, a member of the *Paramore* court martial, was drowned along with the entire crew of over 300 when his ship the 60-gun *Mary* fell apart. The Great Storm was one of the worst natural disasters in the history of the Royal Navy and over 1,500 seamen lost their lives. Once more, so much for Halley's other comment that *'great Shipps are thought unfitting to be ventured at sea for 2/3 of the year'*.

Shovell yet again managed to avoid responsibility. He pointed out that he was only following the orders of Prince George and the Admiralty Board, which was true. However, no one expected him to take so long in carrying out those orders that he would put his entire fleet in jeopardy. He should have followed his own stern advice and brought his fleet home a month earlier, or at least followed the example of his Dutch allies. The Admiral's lucky survival and escape from blame cost him his 8th life.

Others were equally affected by this tempest. Thousands of fishing boats were sunk and lives lost due to falling trees and masonry across a wide swath of southern England. 400 windmills were wrecked, some of which apparently caught fire due to friction caused by the speed of the spinning sails. The newly built Eddystone Lighthouse, constructed to last 100 years and which Halley had so recently admired during his Channel survey, was smashed to smithereens, drowning the keepers and Henry Winstanley its designer.

Even in London considerable structural damage was caused to buildings and some of the Mint's new outbuildings, despite being protected by the Tower of London's outer walls had roofs blown off. Just when Newton was trying to organise tin sales on a massive scale.

This latest interruption to his efforts to re-organise the Mint had been caused by Queen Anne's well-intentioned scheme to pay the Cornish tin miners a fair working wage. She had ordered the Treasury to contract to purchase 1,600 tons per year for seven years at a much improved price of £69.10s per ton (£14,000) [1] and her Mint to prepare it for re-sale world-wide. Agents were appointed in various countries and sales commenced in January 1704.

Immediately merchant ships began loading the heavy casks of tin from the Tower wharf and almost as quickly captains of English warships realised they too could turn a quick profit on their own account by selling tin at foreign ports of call - especially to the two appointed agents in Italy where Cornish tin was in great demand in the production of casting bronze. The use of warships to avoid freight charges was branded unfair by the Levant Company [2] but their protests made matters worse; the Treasury began using Royal Navy ships to transport tin to Italy officially.

On February 24th 1704, Admiral Sir Cloudesley Shovell was made an elder brother of Trinity House, the organisation responsible for maintaining lighthouses round the coasts of the British Isles and devoted to saving the lives of seamen. Shortly after, a goodly portion of the patched up Royal Navy set off to do battle with the French and various Spanish factions, their lordships hopefully having learned a lesson from the previous year's debacle, which had resulted in greater losses through incompetence than those they had managed to inflict on their enemies.

This early season departure caught their protagonists on the hop and the Royal Navy had one of the most successful seasons ever. Gibraltar was captured and a major sea battle was fought off Malaga involving nearly 50,000 seamen and over 7,000 cannons in no fewer than 102 huge wooden battlewagons.

This fight was inconclusive because the Anglo-Dutch force, which had managed to gain the upper hand, had to withdraw many of the English ships through cannonball shortages. Why the shortages? Much ammunition had been expended in battering Gibraltar into submission and no provision had made for replenishment.

By all accounts Shovell, for once actually in an action, acquitted himself brilliantly. The fleet returned home early and was back safely in the Medway by the end of September. In October, Shovell was presented to Queen Anne who gave him a gold snuffbox and he was also feted by the Lord Mayor of London. In November, he was appointed to the Council of the Lord High Admiral and in December promoted Rear Admiral of England. In January 1705, Shovell was given a further promotion; Admiral and Commander-in-Chief of the Fleet. He could climb no higher and still remain at sea yet he was only 54 years old.

The boy from an obscure Norfolk village had come a long long way in those 45 years at sea and it is difficult to imagine him ever swapping his newly won kingship afloat for a life ashore.

While the English had been busy in the Mediterranean, the French West Indies squadron somewhat ineffectively attacked English colonies in the Caribbean. This inflated the asking price for slaves and caused considerable inconvenience to the slave transporters. The result was that the smaller islands petitioned the English parliament for preferential treatment in the supply of slaves because their economies were suffering through labour shortages.

One estimate suggested that the Leeward Islands needed at least 3,200 slave replacements annually. Apart from carelessly losing some of them to the French, what were they doing with their existing stock of (breeding) slaves to need so many just to retain the status quo? Killing them through overwork and providing insufficient medical attention was the straightforward answer.

(1) John Craig, *Newton at the Mint* (Cambridge, 1946), p.57 (2) Ibid., p.58.

CHAPTER 23

Having got themselves properly organised in 1704 and yet again in 1705 the Admiralty reverted to type, lessons unlearned. Shovell's fleet was so late leaving England for the 1706 Mediterranean fighting season that the first of the autumn gales actually delayed their departure!

The plan to attack southern France with land forces had meant that some 10,000 troops and all their equipment (which had been waiting in the Portsmouth area since the spring) had to be loaded into transport ships before joining the battle fleet and this had taken a great deal of organising.

That summer Newton had received instructions from the Treasury to prepare nearly 400 casks of tin (about 80 tons) for delivery to Shovell's fleet. Shovell then left the casks in the naval base at Portsmouth, saddling Rear Admiral Sir Thomas Dykes with the task of delivering the tin to the Queen's agents in Genoa and Livorno [1]. Later whilst ashore in Livorno, Dykes was murdered on the orders of the Governor presumably in revenge for Shovell's previous insulting behaviour. With the death of Dykes, the now obese, famous and exceedingly rich Admiral of the Fleet had used his 9th and last feline life.

Shovell left Portsmouth carrying with him £100,000 (£20 million) in gold and silver coin to pay foreign troops, and probably a great deal more to provide for the victualling costs of his massive armada. When the fleet eventually arrived in the Tagus estuary following a stormy crossing of the Bay of Biscay at the end of October, over 1,000 troops had been lost when some of the transports had sunk or been wrecked, and a further 100 troops had died of disease (or fright).

Shovell sensibly decided to over-winter in Lisbon although this was no great hardship because he was flying his pennant in the 90-gun *Association,* now luxuriously refurbished following the vessel's dismasted and undignified arrival on a Swedish beach after the Great Storm of 1703. He had his two young stepsons with him, one an ordinary seaman and the other a captain's servant. Also on board was a naval captain who was a nephew of Shovell's wife; quite a family party.

When everyone eventually arrived in the Mediterranean in the spring of 1707, complete with complex fighting instructions from Prince George - who still had no experience of fleet actions - they again discovered there was a shortage of ammunition. The fleet and surviving troop transports anchored off the great French naval port of Toulon on the 13th June 1707 with the intention of capturing the city in collaboration with land forces commanded by the Duke of Savoy, who had been waiting impatiently for months. Shovell already knew that in Toulon harbour were 46 French warships unable to put to sea and defend themselves for want of funds and manpower.

Louis XIV was rightly concerned that Shovell would capture or destroy his fleet so he had ordered all but two to be scuttled, hoping to recover them later. The land-based attack on Toulon was a long drawn out failure and Shovell, who had done his best to support the troops with naval bombardments, had to content himself with setting fire to the few French men-of-war whose upper-works still remained above water.

Shovell eventually sailed for home from Gibraltar on 10th October with 11 great ships of the line plus 9 other assorted armed Royal Navy vessels. The fourth-rate 54-gun *Panther,* commanded by Captain Henry Hobart joined the fleet the following day and they all headed post haste for home, the *Association* still carrying a large fortune with her, securely locked in sea chests in the strong room to the fore of the great cabin.

Many years later Horace Walpole was to remind the nation that *'Sir Cloudesley Shovell said that an Admiral would deserve to be broke, who kept great ships out after the end of September, and to be shot if after October.'* [2] Yet here he was, once more 'keeping great ships out after the end of September' and ignoring his own dictate when he could have safely over-wintered either at Buccari now that the harbour had been properly surveyed and deemed suitable by Halley, or in Lisbon again.

As was to be expected they ran into stormy weather once out in the Atlantic, but the winds being generally from the southwest, made their uncomfortable passage speedier. All were looking forward to seeing their homes and families after such long absences. Much to everyone's relief on the 1st November after several days of unusual easterly winds and overcast skies, at least seven fleet navigators did manage to obtain latitudinal back-staff noon sights (open circles figure 8), but the seas were heavy and most failed or did not bother. These sights all put the fleet at or just below the 49th parallel and were in reality between 10' and 30' south of the fleet's true position. As can be seen from the 3½° scatter of plots, none had much idea of their longitude.

The next morning Shovell ordered Sir William Jumper, captain of the 70-gun *Lenox* to take the lead, using the two frigates *Phoenix* & *La Valeur* as scouts. This was the traditional method employed whenever a large fleet was approaching land and Shovell had availed himself of Jumper's expertise in this role in the past. Jumper later gave the time of leaving the fleet as 11.00 a.m., his two lieutenants as 7.0 and 10.0, the captain of *Phoenix* 8.0 and the captain of *La Valeur* 11.0, which underlines the generally sloppy attitude existing at that time in the Royal Navy.

The trio hightailed it off to the northeast presumably searching for land. Later in the day Jumper felt they might be standing into the danger of running into the Scilly Isles or the Lizard and ordered a change of course to southeast. Unfortunately by the time he had decided to alter course his three ships were in fact already to the northwest of the Scilly Isles and now actually behind the main fleet rather than scouting out ahead (figure 8).

Meanwhile the fleet took a more easterly course before most ships heaved-to in the late afternoon to take soundings [3] which indicated they were bordering the 50 fathom line at the entrance to the English Channel.

They then set off again heading slightly south of east, a course which put most of the fleet tragically on a collision course for the deadly outlying ring of granite rocks and hidden reefs off the southwest corner of the Scilly Isles and due to arrive there about 3½ hours after sunset.

As night closed in on the fleet, Shovell must have over-ruled those urging caution. Instead of shortening sail or heaving to until daybreak, the Admiral's flagship actually led the way, presumably relying on meeting up with one of Jumper's frigates if danger lurked. A strong following wind now pushed *Association* headlong on into the oncoming night with the rest of the fleet tagging along behind like a long string of attentive ducklings instinctively following their experienced mother.

At which point Shovell's long run of good luck finally ran out. Lookouts on *Association* failed to see the light of St. Agnes Island's coal burning lighthouse 150 feet above sea level (funded by Trinity House) or the surf breaking on the rocky fangs or surging up and over the hidden reefs, and the great battleship rammed into the infamous Gilstone Ledges at its lumbering full speed of about 4½ knots.

From what little is known of subsequent events it seems most likely that Shovell, his two stepsons and nephew-in-law Captain Edmund Loades, were in the Great Cabin when *Association* struck. The unexpected halt would have thrown bodies, chairs, plates, goblets - and anything else not firmly anchored down - against the strong-room bulkhead in one chaotic heap. Before any of them could disentangle themselves, a huge following wave lifted the now stationary 50-metre long ship high above the reef and the following trough then promptly dropped all 1,600 tons of her back down onto the top of the ledge. The weight of her 90 guns split her seams and the ship simply fell apart, her three massive masts, no longer supported, toppling slowly sideways. The guns, hundreds of tons of ballast, stores and broken chests of coin as well as more than 700 men were dumped unceremoniously into the cold and turbulent water swirling round the Gilstones. One minute the proud and powerful flagship of Queen Anne's Royal Navy and the next, she was quite literally nothing but large pieces of driftwood.

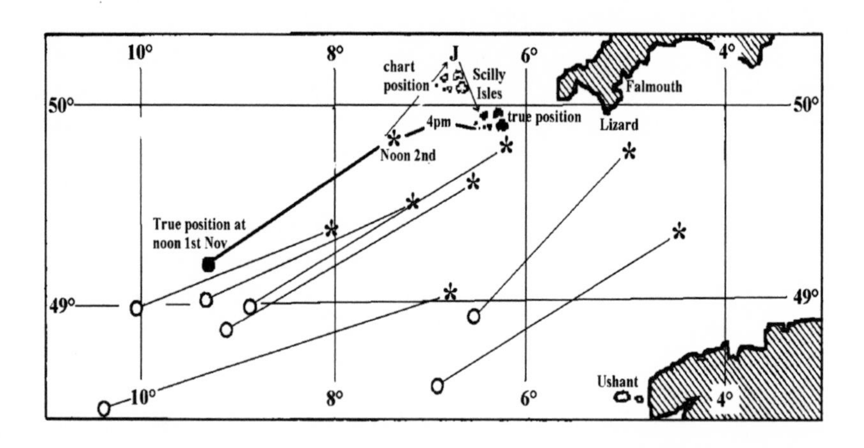

Figure 8. Approximate track of the Shovell fleet on 1st & 2nd November 1707. Open circles are the estimated positions of the 7 ships whose navigators managed to take noon sights on 1st November with extrapolated DR positions (*) *for noon on the 2nd. The Jumper* (**J**) *detachment directions are shown by thin arrows. The main fleet hove to for a short while at about 4pm on the 2nd to take soundings. The true position of the fleet on the 1st November is indicated by the solid circle and subsequent track by the thick line. The erroneous chart position of the Scilly Isles is shown to the northwest of its true position.*

Somehow Shovell, his dog, his two stepsons and Captain Loades finished up in the Admiral's barge. The family party at this point in the disaster was remarkably still intact. The most likely explanation for this miraculous escape from the remnants of *Association* when every other member of the crew of over 700 drowned at or near the disaster site within minutes, is that the stern section of Association was the last to break up and the Admiral's barge having been dislodged from its davits was floating in its lee.

The small fire ship *Firebrand* followed *Association* onto the Gilstone Ledges but drifted off again only to sink shortly after. Only her captain Francis Percy and some 20 of his crew were washed up on St. Agnes Island alive. The 70-gun *Eagle* smashed into the nearby Tearing Ledge, immediately broke up and her entire crew was lost. *Romney* sank somewhere close by although her remains have never been located. The sole survivor from that ship was the Quartermaster George Lawrence who escaped death by clinging to an oar before being swept onto a rocky outcrop off Annet Island from where he was rescued the next morning.

The largest ship in the fleet, the 96-gun *St. George* also ran onto the Gilstone Ledges but was miraculously lifted off again by a great wave before sustaining any lethal damage and all the crew survived.

Admiral Sir Cloudesley Shovell, his relatives and his pet dog somehow got clear of the tangle of spars, sails and rigging that littered the wreck site in the admiral's barge. It did them no good in the end and as the Moon was setting the barge was dashed to pieces on rocks five miles to the east and the bodies, including that of Shovell's pet dog, were washed into Porthelic Bay on St. Mary's Island. When the Admiral's life expired, *Association* too had now been lost with all hands.

Having expended all his nine lives, his luck and that of nearly 2000 sailors who had relied on his commanding judgement had finally run out. Shovell's body was discovered by locals, stripped of its clothing and buried well above the high tide line. Later some rumours suggested he had been alive when found and had been murdered for his jewellery. There is no evidence to confirm either this or the unbelievable claim that he had hanged a sailor the day before the disaster for daring to question his navigational expertise.

The great guns of *Association*, along with the vast treasure she was carrying fell to the seabed. In addition to no less than 10 sea chests packed tight with coin to the *rumoured* value of £3 million (£600 million), *Association* also apparently carried two full dining services in gold and silver. Whatever the truth, there it all now lay on the seabed or in amongst a huge jumble of massive underwater boulders that were constantly pushed this way and that by the powerful surges of the waters of the mighty Atlantic Ocean. Seas so powerful that they could move massive 30-ton rocks and would eventually place one of these tidily on top of a bronze cannon, which Shovell had taken as a prize at Vigo.

But what of *Lenox, La Valeur* and *Phoenix*? *Lenox* and *La Valeur* suddenly found themselves in amongst rocks three miles north of the Shovell disaster but, at 3 a.m., some seven hours later. Miraculously they avoided running into anything, realised where they were and managed to anchor in sheltered water for the night before proceeding to Falmouth next morning, unaware of the earlier disaster to the south of them. Jumper assumed that *Phoenix*, the third ship of his squadron, had somehow also missed hitting anything and was now nearing Falmouth; a very peculiar assumption. In fact Captain Michael Sanson commander of *Phoenix* had seen the light on St. Agnes, but thinking it was a light from one of Shovell's ships, headed towards it to report. A lookout spotted an unlit ship dead ahead in the darkness and they prepared to go alongside in order to discover its identity.

Unfortunately it was a rock and *Phoenix* was holed on an underwater companion. The heavy seas lifted the small ship off and, leaking badly, she anchored in nine fathoms of water less than two miles from where *Lenox* and *La Valeur* were about to drop anchor. At daybreak *Phoenix* fired guns and a local fisherman came aboard and piloted her into a sandy bay on the island of Tresco.

From the original fleet of 21 ships, four were sunk (*Association, Romney, Eagle* and *Firebrand*), and two (*St. George and Phoenix*) were damaged. *Lenox, La Valeur, Griffin, Royal Anne, Somerset, Monmouth, Swiftsure, Torbay, Isabella* and *Orford* all had very narrow escapes. With the single exception of *Panther*, every one of the great ships of the line (including the three commanded by persons sitting in judgement on Edward Harrison), either foundered or could easily have done so.

During the night the storm blew itself out and by morning the skies had cleared and a light breeze from the north created gentle rippling wavelets across the Scillies anchorage. *Panther* and several of the smaller ships at the rear of the fleet had been unaware of the disaster and arrived in Plymouth the next day, wondering what had become of the main fleet. But to the crews of the ships that had been fortunate enough to have avoided the reefs and managed to anchor in the sheltered waters of St. Mary's roadstead, it was soon obvious many ships had been sunk in the night. Broken timbers, tangled cordage and bits of tattered sailcloth were drifting in clumps as far as the eye could see. Tangled amongst this flotsam were the bodies of drowned seamen.

The tide and the gentle breeze were between them now pushing all this mass east towards the shoreline of the island of St. Mary's where men women and children could be seen rushing hither and thither like ants, collecting bits of wood and spars, tables, chairs, wooden buckets and food chests. A few were stripping stranded bodies; time enough to dig pits above the shoreline to bury them in due course.

By long tradition they would not be given Christian burials in one of the tiny churchyards because no one knew if the dead had been of the Christian faith. To some extent this was simply an excuse to avoid the crippling cost of the proper burials of seamen, which local inhabitants were obliged by law to pay for.

Soon all the ships were launching boats and these were being rowed towards the western outer reefs in search of booty first and survivors otherwise, or were landing officers and marines ashore on St. Mary's to gather information and recover Admiralty property before everything was spirited away.

The captain and surviving crew of *Firebrand* were found on the island of St. Agnes, whose beaches were littered with hundreds of bodies, and the Quartermaster of *Romney* was rescued from his rock. Although badly battered and suffering from hypothermia, he was questioned at length. These eyewitnesses convinced everyone that *Association* had been lost with all hands and an intensive search was organised for the Admiral's body. The Quartermaster was abandoned to his fate, cuts unattended despite the presence of at least one ship's surgeon, but fortunately for Lawrence a local man took pity on him and persuaded a civilian doctor to treat his wounds.

It was the purser of *Arundell,* a man-of-war from the Welsh squadron visiting the Scilly Isles at the time of the disaster, who was responsible for finding the hastily dug grave and having the body dug up and identified. The bodies of Edmund Loades and Sir John and James Narbrough were also recovered from Porthelic beach and re-buried in the chancel of Old Town Church, St. Mary's.

In all between 1,800 and 2,000 sailors drowned. The mass graves dug hastily on St. Agnes remained clearly visible for many years and the fact that the Admiralty never bothered to give these brave men and boys a decent burial, or even to record the exact number of deaths, is a sad example of how England treated her fighting seamen in the early 18th century.

When divers were building the base of the first stone Bishop Rock lighthouse in 1851, they lived in a shack on the island of Rosevean, coincidentally the nearest piece of habitable land to the Gilstone Ledges. On days when the weather prevented work out on Bishop Rock they went diving for treasure. What they found was never recorded.

In 1967, the professional Cornish diver Roland Morris began salvage operations at the *Association* wreck site. His team, working in dangerous conditions, recovered three of the bronze cannons and large quantities of coin, mostly silver and mostly tucked into places difficult to access. Very few coins on the open seafloor away from the surging underwater currents close to the Ledges, but plenty of cannonballs, shot, and a number of iron cannons.

It was only when they lifted some of the corroded iron guns that they came across thick wedges of coin packed underneath.

Roland Morris made the following observation in his book *'HMS Colossus'*. *'Now I was certain what had happened. As Association broke apart and the admiral's treasure chests broke open, the hundreds of thousands of coins scattered like leaves in an autumnal forest. The sea-bed would have been carpeted with them three or four deep, just as we had found them under the guns, and once upon a time they would have lain in the silt at the rate of three hundred to every 9 feet by 4 inches* [the area masked by each gun] *of the sea-bed.'* [4] There to await the collection of those 19th century divers on their days off. Only under immovable objects and in inaccessible crevices had coins been left undisturbed.

The initial reaction of the English public to news of the disaster was one of shock-horror. Some 2,000 sailors drowned and but 22 had survived from the four ships that had struck the outer reefs of the Scilly Isles. Their swashbuckling Rear Admiral of England and Admiral and Commander-in-Chief of the Fleet drowned? How could this be? However, it was not long before the more knowledgeable of the populace began asking awkward questions. Most were quietly sidestepped but two key questions could never be brushed under the carpet.

The minor point was that the bodies of Shovell, his personal entourage and his dog were washed up on a beach miles from any other bodies but in the vicinity of the wreckage of his barge. Obviously they had left a sinking ship and the rest of her crew to their fate. However, the really serious issue was neatly summed up in a widely circulated newspaper article a month after the disaster. *'Twas strange that Sir Cloudesley, who was bred to the Sea from his own Infancy, should be guilty of such a Neglect and Mistake, upon our own Coast too!'* [5]

Strange as that may be, stranger events were about to unfold. Standard procedure following the loss of a Royal Navy vessel was (and still is) to convene a board of enquiry or court martial. In this case, Great Britain's Lord High Admiral managed to avoid adverse publicity in a manner any modern politician would envy. Prince George was either not at all stupid or was acting as a convenient mouthpiece for the Admiralty.

On the 21st November 1707, before Shovell's funeral cortège had even reached London from the West Country, Prince George issued the following order to the Navy Board:- *'Gentm. Where as her Majts Ships ye Association Eagle and Romney, were on 22d. of last month* [2nd November] *unhappily lost on the Rocks near the Island of Scilly, and all their officers & Companys drown'd, soe that noe Inquiry can be made Into their loss at a Court Martiall, I doe therefore, in Consideration of the Misfortunes of the Widdows & other Relations of the Officers and*

Companys who perished as aforesaid, hereby desire and direct you to cause to be paid to such persons as shall be empowered to receive the same, the wages due to them, to the time the said ships were lost, without expecting any accounts from their Officers.' [6]

The loss of *Firebrand* whose captain and 20 crew members had survived was not mentioned, and neither was the fact that the boatswain of *Romney* was still alive, just.

In order to pay the dependants of the 2,000 drowned seamen the *back pay* they were owed, each ship's wages book had, by naval regulations to be produced. Some men would have been owed more than a year's wages, but many would have been given an advance by the ship's purser. If these books were missing - and they were - an enquiry of some sort would first have to be convened to establish *why* these books were missing before the men's relatives could receive monies due.

Additionally by implication, should any seaman be found guilty of anything at such an enquiry his relatives may not have been entitled to the back pay. Prince George, pleading hardship on behalf of these relatives, had ordered the Navy Board not to investigate the disaster using the (untrue) excuse of there being no survivors to interview. Thus there was no proper enquiry into the disaster, the reasons for which were well enough understood by the Admiralty.

First, their failure to properly notify all Royal Navy captains and master navigators of Halley's discoveries.

Second, the fact that they had placed an embargo on Newton's angle-measuring device and in so doing had denied any of the fleet navigators access to a far more accurate method of determining *latitude*, especially in heavy weather.

A robust positive response to *either* point would have prevented the disaster. Embarrassingly in both instances their brave but late commander-in-chief was implicated, as to a lesser degree was the Secretary to the Admiralty Josiah Burchett.

A court martial or public enquiry would have forced a cross-examination of Halley over the publication of his broadside warning pointing out that charts and pilot books dangerously gave the latitude of the Scilly Isles as much further north than their true position and exposed the Admiralty's reluctance to spend money on producing its own charts and pilot books.

The moment someone questioned him as to how he was so certain of his facts, the truth would have emerged regarding Newton's instrument and the Admiralty quite rightly held responsible for the deaths of thousands of their men. 30 years later RN captains were still protesting to the Admiralty over the unreliability of charts and navigation manuals.

Then there was the question of magnetic variation. Few navigators had any faith in their magnetic compasses, many of which had not been checked for years [7]. Halley had also publicised his discovery that the magnetic variation had swung westerly by at least 3½° since last properly checked; a fact any competent navigator could have confirmed with an azimuth compass.

None of the fleet navigators appear to have made any allowance for this during their voyage north, but it is very difficult to determine from the surviving log books exactly what effect, if any, this would have had on the fleet's course because so many of the compasses were unreliable anyway. Jumper, in an effort to avoid any responsibility, attributed the disaster to defective compasses and emphasised his point by returning no less than 10 of those carried by his ship *Lennox* to Chatham dockyard *'broke & in pieces'* [7], whist his officers testified that the others still on board were also useless.

Lady Shovell, one of the richest women in England in her own right then attempted to claim a widow's pension from the Admiralty. This was refused because her husband had not lost his life in action. She then kicked up a fuss over two rings allegedly taken from her husband's body before it had been hastily buried above Porthelic Bay, offering a reward for the return of the valuable emerald ring; a reward that was never claimed.

Eighteen months later the Admiralty sent a recovery team to the islands almost certainly looking for *Association's* treasure chests and any other valuables that might have been spirited away by the local population. They were authorised to question the locals *'using fair means or foul.'* Despite apparently being given the right to torture Scillonians, neither treasure nor emerald ring were recovered.

27 years later on 1st February 1735 a local woman by the name of Mary Mumford allegedly made a deathbed confession and the emerald ring was returned to the Shovell family heirs. The huge green stone became the centrepiece of a locket with Cloudesley Shovell's name and date of death engraved on the back; it was last seen in 1884. The second ring was not surrendered and many islanders still believe that it is in the possession of a descendant of Mary Mumford; tradition has it that if this ring ever leaves the islands, they will sink.

A State funeral was arranged for the Commander-in-Chief of Her Majesty's Royal Navy, who was to be buried in Westminster Abbey. Elaborate plans to bring the body by river were vetoed by the Crown on the grounds of cost, and instead a funeral procession of over 100 carriages wound its way through the streets of London to the Abbey by torchlight. This was expected to be watched by tens of thousands wishing to pay their respects to a great naval hero but such was the public outrage that, according to one report only odd knots of curious bystanders turned out.

(1)Roland Morris, *Island Treasure* (London, 1969), pp.122-3.
(2) Simon Harris, *Sir Cloudesley Shovell; Stuart Admiral* (Spellmount, 2001), p.333.
(3) W. May, *The Last Voyage of Sir Clowdisley Shovel* (J.Inst. Nav.,1960),Vol.13/3, pp.324-332.
(4) R. Morris, *HMS Colossus* (London, 1979), p. 200.
(5) S. Harris. p.369.
(6) Ibid & Public Records Office; Admiralty Letter Book. No.18 p.481.
(7) W. May. *Naval Compasses in 1707* (J. Inst. Nav., 1953),Vol. 6/4, pp.405-409

CHAPTER 24

Newton's efficient supervision of England's re-coinage which had commenced in 1696, his secret invention, his election to the presidency of the Royal Society, the publication of *'Opticks'* and his knighthood in 1705 spanned a mere 10 years. The reclusive genius had metamorphosed; no longer the vulnerable grub seeking shadows whenever he was illuminated; he had become a fully-fledged predator. Newton had been bullied at school until goaded once too often which then left his bigger opponent scarred for life when he scraped the bully's face along a wall. He was now in a position to settle a few old scores in a far subtler manner.

In the spring of year of the Great Storm of 1703, Robert Hooke had died a lonely, sadly neglected and generally unloved old man, having been bed-ridden for more than a year. While he lay dying, documents and assets were systematically removed from his room but his thieving 'friends' missed more than £8,000 (about £1.6 million) in cash that was later discovered in a locked trunk under his bed.

Hooke had treated Newton's first beautifully crafted telescope with sarcastic contempt, and instead of welcoming a newcomer had behaved like a spoilt child during Newton's first real foray into the world of scientific experiment. This had tipped the delicately balanced scales of true genius and turned Newton into an unstable recluse. Then, when Newton demonstrated a second wonderful instrument, this time from a position of confidence, he was defamed by his hated rival and frustratingly found himself in no position to defend himself. So his brilliant publication *Opticks* that, on Halley's insistence was going to include a grudging acknowledgement to Hooke, remained unpublished until Hooke's death was announced. Newton then deleted the acknowledgement and one of the greatest comprehensible scientific publications of all time finally saw the light of day and Hooke's contribution was sidelined.

With one of his hated enemies posthumously humiliated the Reverend John Flamsteed, who was at the time, actively engaged in his favourite displacement activity of libelling Halley, now found himself fully in Newton's sights.

John Wallis, one of the founders of the Royal Society and the Savilian Professor of Geometry at Oxford died seven months after Hooke and Halley once more was presented with the opportunity of academic advancement. The Royal Navy captain immediately became the firm favourite for the post even though at the time he was on the Continent on that delicate diplomatic mission on behalf of Queen Anne and was unaware of the vacancy. Within days of Halley's return to England, Flamsteed wrote a letter to Abraham Sharp his some-time assistant and a noted astronomer in his own right. Having already mentioned 'Captain Raymer' in a previous letter to Sharp he now followed this up by another explaining exactly who he was referring to and why and in his second sentence Flamsteed produced another of his customary back-handed compliments. *'Dr. Wallis is dead: Mr. Halley expects his place, who now talks, swears, and drinks brandy like a sea-captain: so that I much fear his own ill behaviour will deprive him of the advantage of this vacancy.'* [1] Flamsteed was wrong and Halley was elected (without any support from Newton).

Having been provided with dwindling amounts of data over the years, it became painfully obvious to Newton that Flamsteed was no longer willing to supply the data demanded, data which were needed to fulfil his self-imposed promise to his Maker. It was equally obvious to Flamsteed that his efforts to fulfil his own undertaking to God would be compromised if he concentrated on meeting Newton's demands.

Immediately following Hooke's death, Newton had managed to get himself elected president of the RS and began bullying fellows into pressuring Flamsteed to publish the data he could otherwise not lay his hands on; Flamsteed was at best unhelpful. Of course he was, what data he did possess, he now knew to be suspect, and of far more significance, his star maps had hardly been started upon; so driven into a corner he prevaricated. Playing on Newton's miserliness, he pointed out that he had already spent nearly £2,000 (£400,000) of his own money running the Greenwich Observatory, part of which had been used to replace the instruments spitefully removed by the RS in 1679. He could not therefore afford to pay to publish his data, a publication that must include *all* his results including his star maps. He knew that if *only* his lunar and planetary data and the technical information on star positions in tabular form were to be published, Newton might use this to solve his 'motions of the Moon' puzzle, (to great acclaim) and his own life's work might never see the light of day.

He did nothing to further his cause when his new president paid another visit to Greenwich soon after *'Opticks'* had been published.

In another letter to Sharp in May 1704, Flamsteed wrote *'My discourse, about the faults of Mr. Newton's* Optics *and correction of my lunar numbers, brought the subtle gentleman down hither on 12th past* [23rd April 1704]. *I thanked him for his book: he said then he hoped I approved it. I told him truly no...'* [2]

At the same meeting Newton was further upset to be told his mathematical predictions of the lunar orbit did not fit the facts retrospectively, especially when it transpired that Flamsteed was quite correct. Something was still badly amiss with his universal gravitation theory. Flamsteed had pointed out that Newton had neglected to check back far enough. As he gleefully reported to Sharp *'...he seemed surprised, and said "It could not be" But when he found that the errors of the tables were in observations made in 1675, 1676, and 1677, he laid hold of the time, and confessed he had not looked so far back: whereas, if his deductions from the laws of gravitation were just they would agree equally in all times.'* [3] Then Flamsteed told Newton that he expected his proposed star maps to cost at least £12 a plate (£300-£350 in total) to produce and he noted that the great man seemed not to take much notice.

Newton must have been thinking furiously about the flaws Flamsteed had just highlighted. Back to the drawing board; surely not? His theory must be right; he simply had not yet managed to obtain the details of all the variables. However, Flamsteed was wrong to think Newton had taken little notice of the plate costs, and he had unwittingly set the scene for Newton's all too public underhand behaviour which, to this day, marks him down as a ruthless tyrant.

The unfortunate Lord High Admiral Prince George was approached for sponsorship of Flamsteed's publication and elected a fellow of the RS. Possibly having a guilty conscience over his role in that disastrous Great Storm debacle the previous winter, he agreed to pay £1,200 towards the printing costs and Flamsteed was given to understand the Prince would also reimburse him the £2,000 that he was out of pocket by. Flamsteed's hopes were raised.

On March 10th 1705, a scant two months before Newton was knighted by Queen Anne, Flamsteed *'met with Mr Newton accidentally at Garraway's: talked with him about the printing, and an honorable recompense for my pains, and £2000 expense...'* [4] He raised the same point several times, once spending *'waterage 1s., horse 6d., coffee 4d.'* [4] in order to meet with Newton and his associates. However Prince George had decided to limit his investment to the £863 (£170,000) *estimated* costs of publication; estimates which apparently made no allowance for the cost of Flamsteed's expensive *Atlas coelestis* star maps but did include the cost of paying someone £100 (£20,000) to check and *correct* Flamsteed's data and for paying two calculators £180 (£36,000) for computing planetary and lunar trajectories.

Flamsteed quite rightly objected to having his data corrected by a third party, especially when he discovered Halley might be asked to oversee the typesetting and proof corrections. He also complained that others might be paid to produce the planetary data Newton was waiting for. Surely Newton, by now a rich man, could pay for this himself, a telling point which, when aired did nothing to improve matters.

Newton retaliated viciously by approving extravagant typesetting costs for the first sections of the catalogue which contained most of the data he was seeking; at the same time confirming there would indeed not be enough money available to print Flamsteed's star maps. It was at about this time that Flamsteed began referring to Sir Isaac as 'SIN'.

The pages for the first section were lifted from the press in 1706 but with the unexpected death of the Prince at the age of 55 in 1708, any last glimmer of increased sponsorship or of Flamsteed's hoped-for £2,000 from that source was extinguished. Prince George was buried quietly in Westminster Abbey and the unfortunate man could at last rest in peace following an unspectacular life on Earth.

The two most disastrous and avoidable accidents not involving enemy action in the entire history of the Royal Navy had occurred during the six short years he had held the post of Lord High Admiral. He had been blamed for not issuing specific orders to bring the fleet home in time to avoid the Great Storm and had been persuaded to be less than honest in order to avoid a very public enquiry into the Shovell disaster.

Obtaining sweet revenge on Hooke and Flamsteed had been relatively easy but dealing with an already dead Admiral of the Fleet presented Newton with an altogether more difficult challenge.

Most scientists of Newton's era were well acquainted with the use of ciphers as a means of establishing priority on discoveries or ideas that required further time to confirm. Those used by Galileo Galilei are good early examples. Galileo's observations of Venus in its gibbous (nearly full) phase in 1610 using his home-made telescope implied the planet was on the far side of the Sun at the time and therefore orbited the Sun rather than the Earth. This would confirm the heretical claim of Nicholas Copernicus, so he announced his discovery in a Latin cipher, intending to unscramble it later when he had more evidence.

Heac immatura a me iam frustra leguntur o.y - These are at present too young to be read by me.

By early 1611 Venus had moved to the near side of the Sun and was exhibiting its crescent phase; supposition confirmed, Galileo revealed the hidden message:-
Cynthiae figuras aemulatur mater amorum. - The mother of love (Venus) imitates the shape of Cynthia (the Moon).

Galileo had used abbreviations (*o.y.* - by me) in his first message to make a perfect fit for the second and this had allowed him to construct dual messages *from the one string of 35 letters*. He must have devoted long hours to the composing of hidden messages that could only be recognised as such by someone with an intimate knowledge of the sender and subject.

Galileo had been making new discoveries about other planets as well, including of course his famous observations of Jupiter and its four orbiting moons. This had prompted Johannes Kepler to suggest that if Earth had one moon and Jupiter four, might not Mars - the planet whose orbit was newly recognised as being sandwiched between those of Earth and Jupiter - have two moons? In the summer of 1610 Galileo sent another disguised note to his correspondents announcing a further discovery. But this time *apparently* he had neither the time nor patience to compose a clever masking sentence so he simply scrambled his message. Now the receiver would at least clearly understand he was being sent a hidden message and may even be tempted to try his hand at deciphering it.

s m a i s m r m i l m e p o e t a l e u m i b u n e n u g t t a u i r a s

But this would have been absolutely impossible to decipher in the 17th century unless one had a very good idea of the theme (the key - chapter 6) which Kepler did have. Anticipating that the message referred to the discovery of two moons orbiting Mars, he brilliantly but incorrectly deciphered it as
Salue umbistineum geminatum Martia proles. - Hail, twin companionship, children of Mars.

In fact Galileo's message concerned Saturn! ***Altissimum planetam tergeminum observavi.*** - I have observed the most distant planet to have triple form.

Although Kepler must have been a little upset by Galileo's 'trick' and very disappointed at the absence of Martian moons, the Saturn observation was at least another new 'watch this space' astronomical revelation.

Surely Galileo really had also devoted an enormous amount of time to constructing this second cipher with its double meaning just in case he could discover two moons orbiting Mars (as well as confirm by further observation the odd shape of Saturn - the planet's rings) later in the autumn of 1610?

Although Mars does indeed possess two moons, Galileo failed to spot them, which is hardly surprising given their minute size. In truth Kepler's solution was the better fit and Galileo had cheated again, this time not by making use of abbreviations but by using two of the 'u's' as 'v's'. A common enough substitution in Latin but one that he had avoided in his earlier construction.

This double-meaning type of cipher that contained a safety net in the event of a flaw being discovered in the ongoing research prior to formal announcement was uncommon. The normal response in such unfortunate circumstances was simply to decline to offer a decipher and hope everyone would forget all about it. Which leads directly to the use of a different type of cipher that, as with both Galileo's examples, could 'cheat' if required.

As mentioned briefly in chapter 8, Newton had early in his career used the pseudonym *'One Holy God',* underlining his Arian belief to others without revealing his name to his fellow heretics. In fact the pseudonym he had used was not *'One Holy God'* but the Latin equivalent; *'Jeova Sanctus Unus'* which was in fact also an enciphered Latinised version of his name *'Isaacus Neuutonus'*, a far from anonymous statement to anyone capable of unscrambling it properly and very dangerous to his career prospects at that time. But Newton had covered his back in so far as the decipher actually read *'Jsaacvs Neuutonus'* which would enable Newton to claim was nothing more than a terrible coincidence if confronted by an enemy; his cipher could very effectively cheat with a little Latin assistance. Also mentioned briefly in chapter 8, Newton may well have incorporated a 'signed' confirmation of his Arian principles within a cipher sent to Leibniz ... i319n… , but again he had covered his back by including it in a long string of letters and numbers.

A classic example of the misuse of Latin in manipulating ciphers to fit facts was one that was well known to everyone in late 17th century England. This one is technically a chronogram; the use of Roman numeral letters within a phrase to reveal a significant date. The *'Lord have mercy upon us'* prayer revealed the date of the great plague. 'LorD haVe MerCy Vpon Vs'.... 50+500+5+1000+100+5+5=1665. This was then updated following the Fire of London in 1666 by changing the spelling of MerCy to MerCIe. In either example again making use of the 'v' substitution.

This last illustration leads on directly to the Shovell monument in Westminster Abbey. Where better for Newton to illustrate his fascination with numerology and his knowledge of ciphers than on Shovell's monument? The obese fool had ignored data obtained by his wonderful invention and consequently drowned 2,000 innocent seamen, and then managed to get himself buried in Westminster Abbey of all places. He had died in his 57th year on the 22nd October (o.s.) 1707, a numerical total of

31. Surely he could construct a hidden signed insulting message within the proposed epitaph linked to such a number? There it would be for future numerologists to discover, cut into a monument situated amongst the tombs of ancient kings and queens in the greatest Abbey in the kingdom.

The famous master carver Grinling Gibbons, known to both Newton and Halley, was commissioned to execute the memorial. Whilst Newton was unlikely to have played any real part in composing the overall text of the epitaph, (set out below) he could easily have influenced the layout of the lines and advised on the use of capital letters. Certainly both are decidedly odd - the capital 'S' in the 5th line is cut in italic script suggesting it may have been altered from lower case at the last minute. The length of lines are peculiar (all of which start with a capital letter but only sometimes end likewise) and the use of 'In' at the beginning of line 12 is possibly not the best choice. The full layout is as follows:-

Sr CLOUDESLEY SHOVELL Knt
Rear Admirall of Great Britain
And Admirall and Commander in Chief of the Fleet
The juft rewards
Of his long and faithfull *S*ervices
He was
Defervedly beloved of his Country
And efteem'd, tho' dreaded by the Enemy
Who had often experienced his Conduct and Courage
Being Shipwreckt
On the Rocks of *Scylly*
In his voyage from Thoulon
The 22d of October 1707 at Night
In the 57th year of his Age
His fate was lamented by all
But Efpecially the
Sea faring part of the *Nation*
To whom he was
A Generous Patron and a worthy Example
His body was flung on the fhoar
And buried with others in the fands
But foon after taken up
Was plac'd under this Monument
Which his *Royal Miftrefs* has cauf'd to be Erected
To Commemorate
His Steady Loyalty and Extraordinary Vertues.

The total of numbers in the text, all relating to age and date of death (2+2+1+7+0+7+5+7) add up to 31. *Deservedly* is the 18th capitalised word, *Shipwreck'd* is the 26th, *Scylly* is the 29th and *In* is the 30th. 1+8+2+6+2+9+3+0 also add up to 31. ***Deservedly Shipwreck'd Scilly. IN.***

The beauty of this construction is that the sculptor need not be aware and the insult was not obvious. The very existence of several less clear-cut alternatives (see appendix 9) could again enable Newton to claim, should he ever have need to, that the insulting enciphered comment was a mere coincidence caused by the inclusion of the word 'in' a couple of times.

(1) Francis Baily, *An Account of the Revd. John Flamsteed* (London, 1835/1966), pp.212, 215, 671, 751. (2) Ibid., p.216. (3) Ibid., p.217. (4) Ibid., p.219.

CHAPTER 25

In 1709, Flamsteed's Royal Society dues were not paid and Newton immediately had his name removed from the membership list [1]. This was a mistake because now the RS had no direct link with or control over Greenwich Observatory. Halley unwittingly solved the problem.

The second (1710) edition of Streets's *Astronomia Carolina* or Caroline Tables (an almanac of solar, planetary, lunar and stellar positions etc. - see chapter 29) included an appendix by Halley which provides a tantalising glimpse into his method of determining longitude at sea utilising lunar data he had previously acquired with observatory instruments. *'....so in the remote Voyages I have since taken to ascertain the* Magnetick Variations, *they have been of signal Use to me, in determining the* Longitude *of my* Ship, *as often as I could get Sight of a near* Transite *of the Moon by a known* Fix'd Star: *And thereby I have frequently corrected my Journal from those errors which were unavoidable in long Sea-Reckonings.'*

Apart from Halley confirming for the first time that he had indeed kept a *Paramore* journal, he mentions for the first time since the Admiralty embargo that he had determined longitude at sea; yet provides no details. He mentions making use of *known* stars, yet names few names. He mentions 'near transits' (appulses) but gives no explanation as to how such observations could be converted into longitudinal positions. He even fails to mention his successful Hyades cluster method (chapters 17, 18, 19 & 21 and appendices 4 & 8).

To have properly explained how he had determined longitude (or latitude) on board *Paramore* was obviously still not permissible in 1710. The best Halley could now do was to include a hint and once again quietly put the Astronomer Royal firmly in his place in relation to his own feeble efforts regarding the longitude by lunars quest, this time without so much as a mention of his name.

Halley's public comments regarding the possibilities of determining longitude by the lunar distance method in a widely used nautical almanac allowed Newton to petition Queen Anne to authorise a committee under his charge to supervise the Astronomer Royal's activities at Greenwich if her late husband's money was to be properly utilised. She agreed and a Board of 'Visitors to the Royal Observatory' was formed, but this underhand behaviour only slowed the delivery of data further, and late in 1711

Flamsteed was summoned to explain why he was not making observations at Greenwich in accordance with the 'Visitors' instructions.

When Halley intercepted him before the meeting and invited him for a dish of coffee possibly hoping to diplomatically smooth the waters in advance, Flamsteed brushed him aside and marched into the lion's den spoiling for a fight with his real adversary. Predictably he was openly insulted by Newton, who first rudely pretended he could not hear what Flamsteed was saying and then when Flamsteed again pointed out he had *'expended above £2000 in instruments and assistance; ... the impetuous man* [Newton] *grew outrageous and said 'We are, then, robbers of your labours?' ...... after which, all he said was in a rage: he called me many hard names; puppy was the most innocent of them.'* Later during the same meeting Flamsteed recalled that Newton *'charged me, with great violence (and repeated it), not to remove any instruments out of the Observatory...'* [2]. Considering the behaviour of the RS in removing their loaned instruments, and that others were Flamsteed's personal property, such a comment was nothing short of malicious.

Immediately following this latest dispute, the publication of the planetary and lunar material (*Historia coelestis - Book II)* was completed and Newton had, for what they were worth, obtained most of Flamsteed's 'ecliptic' data at no cost to himself. Flamsteed was absolutely furious at Halley's efficient involvement, raising any number of objections; some valid, some petty quibbles.

Disapproving of Halley's method of allocating numbers to 'his' stars within each constellation was both petty and ironic. To quote Owen Gingerich; *'Thus the familiar "Flamsteed numbers", which eponymise the First Astronomer Royal for hundreds of amateur astronomers who might never otherwise have heard of him, were actually an* [Edmond Halley] *invention spurned by the ever-proper Revd. John Flamsteed.'* [3] At about this time the old story of Halley having cuckolded Hevelius mysteriously started doing the rounds again soon after Halley's portrait was rather unwisely hung alongside that of Hevelius in Oxford's Bodleian Library.

On 29th June 1712, Halley accompanied by his wife and their three adult children paid a conciliatory visit to the Astronomer Royal. There in front of witnesses Halley offered to burn all the catalogue copies Flamsteed was so bitterly complaining of which had not already been distributed if Flamsteed would only in turn promise to print his own. Flamsteed seems to have considered this offer was a trick to prevent him from publishing and in order to avoid falling into such a trap, refused point blank even to discuss the offer [4]. This must have been an eye-opening experience for Halley's family.

Predictably, Flamsteed's expense claim was never met. Newton had used first Prince George, then Halley and finally Queen Anne to demolish the ailing Astronomer Royal on his behalf. In 1713, a second edition of *Principia* was published but Newton's upgraded lunar theory which occupied Propositions XXV-XXXV of Book III was unconvincing and many thought he had exhausted his powers in his long-running fight with Flamsteed. So after all that effort Newton now had to work out how to persuade others to use his basic mathematical formula to produce the lunar almanac in order to complete his mission. He wasted little time.

Another person who had long been unhappy with progress on the longitude quest front was Thomas Axe who, when he died in 1691 had left £1,000 (£200,000) in his will to any person who should *'make such a perfect discovery how men of mean capacity may find out the Longitude at Sea, soe as thatt they can truely pronounce upon observations if within halfe a degree of the true Longitude'* naming four referees including the Oxford Savilian Professor of Geometry [5]. In 1696 Edward Harrison commented on the prize *('never to be paid I think')* in his publication *'Idea Longitudinis'* which had carried that unfortunate allegation against Halley.

One proposal for the Axe longitude prize, which had been communicated to both Newton and Flamsteed, may have been submitted in 1706, two years after Halley's election to the Savilian chair. This apparently related to an angle-measuring device of unspecified design [6]. Nothing further was heard of this instrument or of the Axe longitude prize despite the considerable sum on offer, and under the terms of the will, that offer eventually lapsed.

In 1708, the Royal African Company found itself facing claims that its slave-trading monopoly was damaging colonial expansion because of its inefficiency over maintaining a regular supply of slaves to the English colonies. Data from a cross-section of colonial Governors' returns highlighted the RAC's plight; in the previous 10 years the company had only supplied 18,000 Negro slaves to them whereas private traders from elsewhere had made up the shortfall by shipping no less than 75,000. Clearly the RAC was no longer entitled to special treatment and the slave market should be legally opened to competition from other English companies.

The RAC countered by pointing out that their policy of buying their own ships and still basing many of them in the Port of London was vital to the future of the City. However, this did not satisfy the powerful provincial commercial groups who wanted the RAC's monopoly broken in order to develop the ports of Bristol and Liverpool and a long battle over free trade ensued. The French were still not helping matters with their annoying habit of raiding English colonial plantations and sailing off with any slaves they could lay their hands on.

Although a reluctant parliament admitted something should be done, no action was taken and suggestions that the Admiralty might consider offering financial incentives were ignored.

At which point the swashbuckling William Dampier sailed into the story once more. Following his court martial, the publication of the first part of his second successful book *'Voyage to New Holland'* had brought him to the notice of Queen Anne. She questioned him closely on his exploits and then gave approval for a two-ship privateering expedition into the Pacific via Cape Horn, which set off in 1703.

This venture was such a disaster that one of the crew (Alexander Selkirk) was abandoned on the eastern Pacific island of Juan Fernandez at his own request! Crews mutinied, both ships sank and Dampier found himself imprisoned in the Dutch East Indies. He escaped and eventually returned home three years later.

Undaunted he set out again in 1708 but this time in the role of a pilot rather than an expedition leader. They collected Selkirk from Juan Fernandez, beat up the Spanish along the length the western seaboard of South America and sailed in triumph up the Thames in 1711 with prizes and booty valued at £170,000 (£34 million).

Dampier promptly published another best seller and his adventures were later used as a basis for Samuel Taylor Coleridge's poem *'The Ancient Mariner'* and the Selkirk incident was converted into Daniel Defoe's *'The life and adventures of Robinson Crusoe.'* Dampier's well-publicised return could not have been timed better; the South Sea Company had just been formed.

This company had been established with the intention of trading with Spanish America on the anticipation of being granted a monopoly by the English parliament. The ratification of the Treaty of Utrecht curtailed Spanish slave-trading activities and amongst other things (such as ceding Gibraltar to Britain officially and letting the Royal Navy have their cannon balls back) granted Britain world-exclusive rights to sell West African Negro slaves to Spanish colonies in the New World. The SSC was in business and the RAC's monopoly finally broken.

Influential SSC stockholders now added their weight to the protests of ship owners and marine insurers, lobbying their members of parliament for a general review of outdated marine navigation methods and for help with encouraging the quest for longitude, often citing the recent Shovell disaster to support their arguments. Even today, in many people's minds this infamous accident is erroneously linked with the inability to assess *longitude* at sea, partly because the real reasons for the tragic losses were never made public.

In the summer of 1713, William Whiston and Humphrey Ditton published a note in *The Guardian* newspaper setting out their ideas for solving the longitude problem. Later in the year, they repeated their claim in *The Englishman* but following Queen Anne's speech at the opening of parliament on 13th March 1714, they saw a real opportunity to advance their case. The Queen had included in her speech the remark, *"Our situation points out to us our true Interest; for this Country can flourish only by Trade; and will be most formidable by the right application of our Naval Force"* [7] .

Whiston and Ditton hastily submitted the details of their scheme to parliament, at the same time urging the passing of a *'Bill or Clause of a Bill....to Appoint a suitable Reward for such as shall first lay before the Publick any sure Method for the Discovery of that LONGITUDE...'* [7]. They then organised a petition from a high-powered delegation of Captains of Her Majesty's Ships, Merchants of London and Commanders of Merchantmen which forcefully pointed out that *'The discovery of Longitude is of such Consequence to Great Britain, for the safety of the Navy, Merchant Ships, as well as Improvement in Trade, that for want thereof, many ships have been retarded in their Voyages, and many lost, but if due Encouragement were proposed by the Publick for such as shall discover the same, some Persons would offer themselves to prove the same, before the most proper Judges.'* [8]

Parliament, many of whose members could see themselves profiting from any such discovery through their newly acquired (or promised) SSC stockholdings, needed no further prompting and a select committee to study the best way to stimulate scientific interest in the quest was immediately set up. Among the members of this committee was Lord James Stanhope editor of *The Guardian* and close friend of Whiston, but as none of the members of this committee had the foggiest notion of how to navigate, they sensibly called on experts for advice.

Exactly how they decided on whom to consult is not recorded, but the choice strongly suggests that they asked the RS, which in 1714 was in effect Newton. Certainly they sought the advice of Sir Isaac Newton and three of his very close associates, Edmond Halley, FRS, Savilian Professor of Geometry at Oxford, Roger Cotes, FRS, Plumian Professor of

Mathematics at Cambridge (on Newton's recommendation), and the philosopher, Newtonian disciple and fellow Arian, Samuel Clarke. Flamsteed, although not invited, at this point decided to invest most of his wife's dowry in the slave-trading SSC. Wren who had not been enamoured of Newton's ruthless treatment of Flamsteed was not a member of this expert panel of advisors either.

But who were Whiston and Ditton and what was the solution they were so keen to promote? William Whiston had been Newton's assistant and had succeeded him to the Lucasian chair of Mathematics until forcibly removed for stating his Arian religious opinions openly. Humphrey Ditton was a highly qualified teacher of mathematics at Christ's Hospital School where Flamsteed occasionally taught navigation. Two very intelligent men, yet the scheme they were proposing was a complete nonsense!

Ships were to be anchored at regular intervals all along the main oceanic trade routes and were to fire star shells at local midnight set to explode with an enormous bang at a pre-determined height of precisely 6,440 feet (nearly two kilometres). Thus navigators of vessels passing within 100 miles would be able to assess their exact position by working out their distance from the anchored ship by comparing the time difference between the sight and sound of the explosion.

All this depended on knowing the precise location of the anchored vessel, which, so the pair suggested, could be determined either by the lunar distance method, the Jovian moons method or by exact timing of lunar eclipses. But if these methods were suitable for determining the position of several hundred wildly rocking vessels anchored on the high seas, the problem would of course already have been solved; as the pair must have known full well.

Apart from these inconvenient flaws, ships could not be anchored in thousands of metres of water and the cost of building and crewing hundreds if not thousands of rocket-firing ships would nigh on bankrupt the nation. Such vessels would also be vulnerable to being boarded by the enemy who could then set off the rockets at the wrong time.

All four experts were close friends of Whiston and must also have known that this idea was an entirely unworkable harebrained proposal. Nevertheless, Coates (who had aided Newton in the preparation of *Principia's* second edition) was in favour, and so, after some apparent hesitation was Newton himself. Halley, ever the diplomat, thought the proposal should be examined experimentally before giving an opinion. The parliamentary select committee, taking their cue from the experts, very publicly approved the ridiculous scheme for a *possible* award.

Between the date of the Whiston/Ditton petition and the date when the longitude prize bill was accepted by the House of Lords, only a little over two months had passed and on 20th July 1714 Queen Anne signed the Parliamentary Act (appendix 7). This offered rewards that, according to the small print, the Treasurer of the Navy was responsible for paying *providing the Navy did not require the funds themselves*; a brilliant touch which, if the Queen had been consulted on the text, suggests she may have advised her late husband over the wording which had neatly side-stepped a public enquiry/court martial hearing into the Shovell disaster.

Would anyone in his or her right mind invest in a State lottery ticket if the small print stated that the jackpot prize would only be paid out if the money was not needed for some other government project? And make no mistake, people were persuaded to invest considerable sums in their efforts to win the prize, which amounted to a hefty £20,000 (£4 million). A careful reading of the rules would also have highlighted the fact that the top prize would have to be shared, unless some entirely new and hitherto unimagined method was discovered.

One stipulation was that the instrument or method was to be subjected to a test voyage to the West Indies and had to be able to indicate the longitude of the port of arrival within specified limits of accuracy precisely as recommended by Newton (see below). Most importantly the method, whatever it was, had to be *practicable* and *useful at sea*. Smaller amounts were set aside to pay for experimental research and development projects.

At this point in the story, Queen Anne died and was buried alongside her beloved consort Prince George; *'An Act for providing a Publick Reward for such Persons as shall discover the Longitude at Sea'* had only just been placed on the statute books. Her distant cousin George Lewis, Elector of Hanover became George I, King of Great Britain, France and Ireland.

The newly appointed Board of Longitude (BOL) which included among its members Newton, Halley, Coates and, much to Newton's annoyance the Astronomer Royal, (for full list of members see appendix 7) then sidelined the 'anchored ships' Whiston/Ditton application that had been used as a basis for encouraging parliament to pass the Act; a parliament that had needed no urging given that the prize on offer would almost certainly make no direct demands on their coffers.

The Board also decided that until some method of stabilising a telescope on board ship could be devised they could not entertain any award scheme using eclipses of the Jovian moons. So much for Flamsteed's written statement of 30 years earlier that mariners could already use this method in conjunction with his tables unless *'their Ignorance, Sloth, Covetousness, or Ill-nature, forbid them...'*

Having been instrumental in approving the Whiston/Ditton scheme, Newton had also been asked for his advice on the wording of the rules governing the international prize and the method he thought most likely to win. He is on public record as informing the parliamentary committee that the lunar distance method was already *'exact enough to determine her Longitude within Two or Three Degrees, but not within a Degree.'* [9] In the draft of a letter written later, he admitted that he had personally (and successfully) advised on the wording of the Act. *'Upon my representing that the Longitude might be found by the motion of the Moon without error above two or three degrees, and that if it could be found to a degree it would be useful if to 2/3 of a degree it would be more useful, if to 1/2 a degree it would be as much as could be desired the Committee of Parliament set premiums upon finding it to these degrees of exactness, and thereby the Act of Parliament points at the finding it by the Moons motion.'* [10]

In other words, after all the nonsense over the anchored ships proposal which everyone directly involved (except possibly Halley) had promoted in order to get a massive longitude prize up and running, Newton admitted that the act of parliament he had helped to establish would stimulate further research into the lunar orbit. And although he was absolutely right in stating that the *'Moon's motions'* were the key, he was careful not to try to explain why in that case he had backed Whiston and Ditton! In keeping with the Admiralty blackout, he was very careful not to mention the problem of accurate angle-measurement on the high seas either. And of course, he was very very careful not to reveal that more lunar data might equate to personal fame when someone (preferably himself) matched the lunar orbit to *Principia's* proposition and produced the all-important lunar almanac.

In truth Newton's angle-measuring invention, together with Halley's data and skill had already just about 'won' the prize 14 years earlier if one ignored the 'practical' clause, a clause Newton had himself approved of. They both knew their experiment had been used to check the feasibility of the measuring device and of the lunar occultation method. Determining longitude once a month or in the years when the Hyades cluster was conveniently placed (appendix 8) was not a solution and this method was of no more use to the ordinary mariner than John Harrison's wonderful chronometer would be (Epilogue). Which is why Newton had publicly and correctly stated that the lunar distance method was not yet accurate enough.

However, no one seems to have questioned his claim that the *'motions of the Moon'* method was already good enough to assess longitude to within two or three degrees. Who had managed that feat and how? Halley using Newton's instrument was the *only person* to have determined longitude by this method, yet those details, which had been hinted at in the 1710 edition of Streete's *Caroline Tables* and which would have shown just how successful Halley had been (appendix 4) were never made public. If they had been, there would surely have been a public outcry over first, the Shovell disaster and then the wording of the act, which appeared tailor-made for Halley and Newton to win the huge prize, whereas it was in truth simply the latest attempt by Newton to obtain more lunar data and confirm his proposition.

Nevertheless several fellows of the RS believed that Halley, sponsored by Newton was preparing to lay claim to the prize; a quite reasonable supposition in the circumstances.

Rumours surrounding Halley's involvement in an attempt on the longitude prize had first reached Falmsteed's flapping ears barely a month after the longitude act had been passed and in a complicated letter to Sharp [11] he commented on a Dr Thoresby visiting his observatory in his absence and advising his assistant *'that Raymers sets up for a finder of the longitude'* , and went on to write *'Tis more than I hear of; but, like enough, that boaster may do so among his clans.'* He then begged Sharp to obtain from Ralph Thoresby FRS, a Leeds topographer, the basis of this information. In such a manner do rumours spread like wildfire.

(1) Francis Baily, *An Account of the Revd. John Flamsteed* (London, 1835/1966), p.90.
(2) Michael White, *Isaac Newton; The Last Sorcerer* (London, 1998), pp. 321-2.
(3) Owen Gingerich, *A unique copy of Flamsteed's Historia Coelestis/Flamsteed's Stars* (London, 1997), p.197.
(4) F. Baily, p.98.
(5) A.Turner in *The Quest for Longitude* (Harvard, 1996), p.129. (6) Ibid., p.122. (7) Ibid.,p.128.
(8) J. Williams. *From Sails to Satellites* (Oxford, 1992), p. 80.
(9) William Andrewes in *The Quest for Longitude* (Harvard, 1996), p.190.
(10) R. Hall & L. Tilling, eds., *The Correspondence of Isaac Newton* (Cambridge, 1977), Vol. 7, pp.330-332.
(11) F. Baily, p.310.

CHAPTER 26

One of the early claims on the Queen Anne Longitude prize had come from an unexpected source; Sir Christopher Wren, founding member and third Royal Society president. The man responsible for persuading his childhood playmate Charles II to issue instructions for building Greenwich Observatory specifically in order to solve the longitude at sea problem. The man who knew as much as anyone about the practical snags attached to this quest. A man who was, at the age of 82 still in possession of one of the finest mathematical brains in the country and known for his religious orthodoxy.

His son Christopher personally delivered a sealed packet to the RS on behalf of his father who was no longer as mobile as once he had been. The packet was addressed to Sir Isaac Newton and came with a signed explanatory letter dated 30th November (o.s) 1714. *'I Present to the Royal Society, a Description of three distinct Instruments, proper (as I conceive) for Discovering the Longitude at Sea: They are describ'd in Cypher, and I desire you would, for Ascertaining the Inventions to the Rightfull Author, Preserve them among the Memorials of the Society, which in due time shall be fully explain'd by yr Obet, Humble Sert.*
Chr: Wren.' [1]

In other words, Wren was not actually making a claim on the prize, merely establishing his position, possibly with the intention of benefiting his heirs. What exactly was in the mysterious package? Now in the 21st century no one can be sure of anything other than that it contained a sheet of paper on which were written three separate lines of capital letters.

However when this document was auctioned with other of Newton's papers inherited by Viscount Lymington, a descendant of Catherine (the daughter of Newton's stepsister Hannah) and her husband John Conduitt at Southeby's in 1936, it was certified as being by Edmond Halley's hand, not Wren's.

This was clearly a copy which Newton had permitted Halley (in his current capacity as Secretary to the society possibly) to make and had then reclaimed. It was also this copy that had previously been cited by Newton's biographer Sir David Brewster in 1855 when he had been given access to these papers. If the original was lodged with RS records as Wren had requested, it has disappeared, as now has the Halley copy.

The three enciphered descriptions were contained within the following three strings of letters:-

OZVCVAYINIXDNCVOCWEDCNMALNABECIRTEWNGRAMHHCCAW
ZEIYEINOIEBIVTXESCIOCPSDEDMNANHSEEPRPIWHDRAEHHXCIF
EZKAVEBIMOXRFCSLCEEDHWMGNNIVEOMREWWERRCSHEPCIP

According to one account these were later deciphered by a Francis Williams of Chigwell in Essex but the first formally published solution appears to have come from Bancroft H. Brown of Dartmouth College, New Hampshire in 1927 [2]. By reversing the letters in each line and omitting every third letter the cipher then reads:-

WACH MAGNETIC BALANCE WOVND IN VACVO
FIX HEAD HIPPES HANDS POISE TVBE ON EYE
PIPE SCREWE MOVING WHEELS FROM BEAKE

The omitted letters spelling CHR WREN MDCCXIV Z (or CHR WREN MDCCXIIII Z in the 2nd string).

Brown's vague interpretation of these three sentences merged the second and third ciphers, suggesting the cipher referred to two instruments not three.

In 2002 Lisa Jardine published a virtually identical interpretation [3] :-

WACH MAGNETIC BALANCE WOVND IN VACVO
FIX HEAD HIPPES HANDS POISE TUBE ON EYE
PIPE SCREWE MOVING WHEELS FROM BEAKE

However Jardine managed a much better explanation than Brown, neatly describing three instruments in modern English. 'WACH MAGNETIC BALANCE WOVND IN VACVO' related to the direct method of carrying standard time by marine chronometer and harked back to 1662 at the RS when *'Dr Wren proposed to try a watch in Mr Boyle's (vacuum pump) engine.'*

The 'FIX HEAD HIPPES HANDS POISE TUBE ON EYE' described, not a juggler's routine but a method for taking accurate angle measurements at sea with a special type of telescope, again relating to research started in 1662. The final 'PIPE SCREWE MOVING WHEELS FROM BEAKE' suggested a device for measuring the speed of a ship through the water via a pipe set in the prow (the beake) [4].

Neither the Brown nor Jardine solutions were precise deciphers, the 'V' at the beginning of the word 'VACVO' was actually a 'Y' (the 7th letter in the first line of Wren's cipher) and the word HANDS should have been spelt HANDES. Jardine had also changed the V in TVBE to a U.

Wren had worked on these three ideas, often in collaboration with Hooke over 50 years earlier (chapter 3) and none had proved satisfactory at the time; or since. The description of the timer had been vague, that of the telescope device already published in some detail in 1705 and known by Wren to be flawed, and the ship's log (if that is what it was) which could not make allowance for surface currents, was of no more practical use than the existing log-line method. At first glance, a set of claims Wren knew were impractical, the crudely enciphered version of which he must have known would be unravelled by Newton in a matter of minutes.

In the circumstances, what exactly did Wren have in mind? If he was establishing a claim, why write to the late Robert Hooke's bitter enemy; why not submit a clearly written claim addressed to Burchett at the Admiralty, to each individual Longitude Board member (including Newton) or to RS Secretary Halley?

If he was intending to submit a holding claim whilst he completed experiments at the age of 82 after a 50-year interlude, why use a crude cipher when he had previously been happy enough to match a Huygens 41 letter affair (a5, b2, c2, d1, e4, h1, i2, l3, m3, n1, o3, p2, r2, s3, t2, u4, y1) with a straightforward 100 letter unbreakable (?) cipher of his own (a7, b4, c4, d4, e12, f1, g5, h6, i10, l7, m1, n7, o9, p2, r6, s2, t7, v4, x1, y1)? This coincidentally was cited by the president, Sir Aaron Klug at an anniversary meeting of the RS on of all days, the 30th November (1999) [5]. If not for his son's comments (see below), one might conclude Wren was in an advanced stage of senility.

Wren had used IIII instead of IV in the second line, the three 'Z's, and the extra letter in HANDES to make each of the lines the correct length in order to contain the full messages, messages which could be easily deciphered; yet he had been forced to degrade his actual descriptions to comply. And why not spell WATCH correctly? The first string would have remained completely decipherable if he had done so:-

OVZCAVYNIIDXNVCOWCECDNAMLANBCEITRENWGARMHHCTCAW
=
WATCH MAGNETIC BALANCE WOVND IN YACVO - CHR WREN MDCCXIV Z

instead of

OZVCVAYINIXDNCVOCWEDCNMALNABECIRTEWNGRAMHHCCAW
=
WACH MAGNETIC BALANCE WOVND IN YACVO - CHR WREN MDCCXIV Z

Wren's reluctance to become involved in unseemly disputes was well known, and Newton must have realised these crude, badly phrased claims masked something much more important. Newton knew that Wren was aware of his own use of complex ciphers to conceal critical passages in his correspondence with Gottfried Leibniz back in the summer of 1676 (chapters 8 and 24) and would never have wasted his time sending Newton such an elementary cipher. Highly intelligent men did not correspond thus.

Having permitted RS Secretary Halley to take a copy and then solved the simple ciphers, Newton would surely have run a check on the three lines using the numerology cipher *if* he, or someone acting on his behalf, had used this type of cipher on the Shovell monument in Westminster Abbey. After all Wren had held the post of Surveyor of Westminster Abbey since 1698 and much earlier had begun employing Grinling Gibbons on a regular basis.

Might word have got out? By this point Newton's nimble brain would have assumed that peculiarities in the original three lines of letters and in the masking solution might well provide the clues. The first was the reverse method of writing employed to make the basic simple cipher a little less obvious and the second was the triple use of the year:- MDCCXIV (1714);1+7+1+4 = 13 (or 39 as the year was repeated three times).

Counting backwards from the final letter 'P' every 13 letters until you have a string of 39 letters produces
WHIHSIIANNPED - BEHCERLDIREEA - NSYGAXCMEVRAE.

At first glance at best another cipher but there are partial words that would have immediately caught Newton's eye in the first batch of 13 letters. WHIS (Whiston) IN (Himself) ANNE (The late Queen Anne) and PED or PAED (possibly paedophilia). Certainly enough to gain his immediate and full attention to a cipher, which surely he would then have unscrambled.

IN WHIS PAED BEH'LD IRE CER(T)AIN EV'RY'N CHASE AXE. G.M.

(To) Isaac Newton. William Whiston involved in paedophilia. Be aware that everyone will be after his blood. Graviora manet (worse to follow). There is an even more sinister interpretation especially as there was otherwise no need to include a reference to Newton in the cipher because it was sent privately to him.

The longer Newton stared at the 39-letter message he had extracted, the more concerned and puzzled he would have become. Wren had found out about the Shovell cipher, yet what else was he trying to say; what was he demanding? Newton could hardy ask the man as this would confirm he possessed the Shovell key.

If he was expected to drop Whiston there was no problem on that score but if he did so immediately and publicly this too would confirm to Wren that he possessed the Shovell key. If he was being told to 'smarten up his act' regarding his treatment of Flamsteed or Leibniz (see below) or to tone down his overbearing behaviour as president of the RS, he faced the same problem. He could take note to prevent Wren (or possibly his son in whom Wren might have confided) from revealing all; but it would have to be in his own good time.

Newton surely would also have realised there was a second meaning to the final 11 letters of the cipher. 'CHASE AXE G.M.' - Go for the longitude prize (which had superseded the Thomas Axe prize) and worse will follow; the skeletons in your cupboard will be revealed.' This was probably what all the fuss was about, there had been plenty of ugly unfounded rumours floating around the coffee houses recently, implying that he and Halley had set up the longitude prize for their own benefit, stirred by his archenemy Flamsteed if the reports were accurate.

Whilst most would have forgotten the occasion when *'Mr Newton shewed a new instrument contrived by him for observing the moon & Starrs, the Longitude at Sea, being the old Instrument mended of some faults, with which notwithstanding Mr Hally had found the Longitude better than the Seamen by other means'* (chapter 18), Wren was hardly likely to have done so, given that his colleague Hooke had subsequently claimed prior invention. And did Wren know of his early forays into the world of alchemy, his injudicious use of the 'One Holy God' pseudonym and his Arian beliefs?

The classic beauty of Wren's underlying threatening message was that it made use of the 'faults' in the overlay ciphers. The odd omission of the 'T' from 'WATCH' is still missing in 'CERTAIN' (but see below); both the letters in the misspelt 'YACVO' are made correct use of; the substitution of the three U's by Latin V's in the smokescreen message provide a clue to a Latin content (G.M.) or perhaps a warning that others were aware of Newton's similar usage in his own pseudonym. Each 13 letter section spells out part of the message in sequence, carrying over three letters from the first section (HIN) and two from the second (HE). Every single one of the 'peculiarities' in the three strings which provided the clues to a successful decipher (but only to a holder of the numeric key) were also essential features in the construction.

Even the standard abbreviated ending to Wren's accompanying letter provided a clue. Finally the omission of the 'T' from that first line would have confirmed to Newton that the threat was genuine and not a ghastly statistical fluke of his own imagining. This is because this letter *had* to be omitted to prevent the useless 11 letter repeat sequence of the 39 letters:-

WHISIIMNVO/WHISIIMNVO/WHISIIMNVO/WHISII

The beauty of the message at a personal level is that it was non-confrontational and although possibly libellous, totally secure. Wren had provided a covering innocuous cipher and only Newton would possess the key to the inner message which he dare not reveal. But why would Newton then reclaim Halley's copy? Either Halley had also managed to extract the hidden message or, more likely was in danger of doing so, as he too held the Shovell key because Newton had shared his 'joke' with the one other person who *knew* Shovell was directly responsible for the deaths of all those innocent seamen.

But is there any evidence to suggest that Newton successfully deciphered the message and acted upon Wren's warnings? Did Newton decipher the message yet Halley fail to do so? As far as is known Newton never did place Wren's ciphers on record as requested, and Wren never did reveal all, although he lived another eight years and according to his son *'the Vigour of his Mind...continued with a Vivacity rarely found at that age, till within a few Days of his Dissolution.'* ; both points suggesting the basic cipher was indeed only a smokescreen.

As for any positive reaction from Newton, he began regularly attending meetings of the commission which supervised the running of Wren's recently completed St. Paul's Cathedral and took a little pressure off Wren in doing so. An odd thing to do given that the magnificent new cathedral represented everything Newton secretly strongly opposed. He also seems to have quietened down considerably and the affairs of the RS became less acrimonious, but this did not happen overnight.

Most of the January/February 1715 *Philosophical Transactions* of the RS was taken up by an anonymous article outlining once again Newton's biased and unfair opinion of his long-running dispute with Leibniz over alleged theft of Newton's calculus ideas; [6] but at least Newton's name was not on the masthead.

But what of William Whiston? Newton may have a withdrawn a little from his very close association with Whiston before he received Wren's package, although it was the Whiston/Ditton seriously flawed longitude solution that Newton had subsequently used to push through the Queen Anne prize earlier in the year.

According to Whiston, for 20 years after their first meeting in 1694, he enjoyed the great man's favour. Then at about the end of 1714 *'...he*

[Newton] *then perceiving that I could not do as his other darling Friends did, that is, learn of him, without contradicting him, when I differed in Opinion from him, he could not, in his old Age, bear such Contradiction; and so he was afraid of me the last thirteen Years of his Life.'* [8] The all-powerful Newton is hardly likely to have been *'afraid'* of a long-term friend merely because he begged to differ.

Whiston then suddenly became friendlier with Flamsteed. In April and again in October 1715 Whiston met with Flamsteed and kept him in touch with the latest developments on the Halley/Newton front [9] and years later would provide advice and help to Margaret Flamsteed, assisting her in the publication of her late husband's precious star maps. This all seems to confirm that Newton had quietly severed all ties with his acolyte soon after receiving the Wren cipher.

In May 1715 Whiston sent a notice to all the members of the Board of Longitude announcing that on 26th and 28th of that month he proposed firing mortar rockets upwards from Hampstead Heath in North London to test his timing idea and requested that *'all Curious Persons within Sixty Miles Distance will please to Observe the same...'* [7]. The only person who actually reported the results seems to have been the Astronomer Royal whose notes were forwarded to other Board members and nothing more was heard of the experiment. In 1716 when Whiston applied for membership of the RS, the application was somehow 'lost'. Then when Hans Sloane (the next RS president) and Halley between them decided to propose and second his membership, Newton stated bluntly that he would very publicly resign if Whiston were to be granted a fellowship. This suggests Halley had either not successfully deciphered Wren's warning or had ignored it, as it was not directed at him.

Was Whiston involved in paedophiliac behaviour? There is no evidence of this but earlier in life he could have been involved in homosexual activities, and if so Newton might also have been involved as there are some odd stories relating to Whiston's nephew whom Newton had employed in his household as a favour to Whiston. The young man allegedly stole 100 guineas (£22,000) from Newton's locked desk after lifting the key from Newton's breeches, presumably not while Newton was wearing them.

Why should Wren hint at paedophilia rather than homosexuality? One could presume that a nod in that general direction was a sufficient warning and the difficulties of constructing this cipher were enormously technical and Wren may have found 'sodmy' or some-such that one step too far and worked on the easier 'paed'.

'Worse' in any shape or form does not seem to have followed, but why should it have? No claim was made on the longitude prize because neither Newton nor Halley had any intention of making an attempt. Newton had not yet solved the lunar orbit puzzle and the angle-measuring device was still on the secret list. Newton had done all Wren could have been expecting yet without admitting anything, and the skeletons in Newton's cupboard were never revealed in his lifetime. Wren's actions might have pulled the waning credibility of the RS back from the brink, but whether they had an inhibiting effect on Newton's scientific productivity we will never know.

A very disappointed Newton soon discovered that his carefully laid plan had not encouraged anyone to produce new lunar data and despite at least 26 pamphlets containing all manner of suggestions being published (and presumably rejected out of hand by Newton) in the first 12 years of its existence, the BOL did not even convene to consider proposals until after Newton's death.

One that was later submitted to Halley involved a 'new' invention (an inferior version of Newton's secret angle-measuring device) and resulted in a long-running sterile dispute that, to this day, rumbles on (chapter 29). Other proposals were rejected on the personal advice of Newton because they related to timing devices. Indeed one of the very early proposals lodged with the Admiralty and passed on to Newton by Burchett in October 1714 involved a 'Shipwatch'.

An unfrocked clergyman by the name of Digby Bull was suggesting to the Admiralty that his special timing device would solve the longitude problem, at the same time warning everyone that the Day of Judgement was at hand. What Newton thought of this application is not on record, although he apparently wrote a brusque 'thumb's down' to Burchett.

In 1721 Newton was still corresponding with Burchett. His latest reply included the following paragraph; *'And I have told you oftener than once that it is not to be found by Clock-work alone. Clockwork may be subservient to Astronomy but without Astronomy the longitude is not to be found. Exact instruments for keeping time can be usefull only for keeping the Longitude while you have it. If it be once lost it cannot be found again by such Instruments. Nothing but Astronomy is sufficient for this purpose. But if you are unwilling to meddle with Astronomy - the only right method and the method pointed at by the Act of Parliament - I am unwilling to meddle with any other method than the right one.'* [10]

Newton was of course, absolutely correct on two points. Longitude could not, and never would be 'found by Clock-work alone' and astronomy was an essential ingredient.

But why the overt rudeness in an official letter to the Secretary to the Admiralty? Certainly he could have been none too happy over Burchett's role in arranging that meeting at the Mint all those years ago. Had the Admiralty now taken his revolutionary instrument into their safekeeping to avoid the secret of its construction falling into enemy hands, as they were to do with John Harrison's chronometer 40 years later?

Newton knew from the outset that the BOL would never have to ask the Admiralty to pay out the *full* prize to anyone unless the Act was amended (Epilogue). The Treasurer to the Navy would never have to make the embarrassing decision to evoke his 'golden share' option by suddenly discovering that the Navy urgently needed the money themselves even if someone succeeded in publishing a lunar almanac or invented a reliable, cheap to manufacture marine chronometer. In either event at most one half of the prize might have to be paid out because longitude (and latitude) could still not be ascertained at sea without the assistance of a fool-proof angle-measuring device to determine local (noon) time, a device which had already been thoroughly proven and which was now a national secret.

It is impossible to reach any conclusion other than that Queen Anne's last Act of Parliament was intended to be nothing more than a political sop to the rich merchants, to quieten critics of an inept and outdated Admiralty, as a publicity gimmick to boost the stock values of the SSC, the new company everyone (including stockholder Newton) hoped would make them all rich and last but by no means least, in order to encourage others to assist a brilliant scientist to complete his holy mission. A deception through and through.

In theory, there had been no need for a longitude prize. Britain already possessed a research institute dedicated to solving the longitude at sea problem; Greenwich Observatory. Also in theory, should parliament at any time consider the progress of research regarding the production of the lunar almanac too slow, they should have channelled funding and fresh brains in the direction of the observatory. Yet apart from authorising a committee to oversee Flamsteed's work which in reality brought his lunar research to a virtual standstill, they had done nothing. They were not even prepared to fund the publication of the observatory's results.

(1) R. Hall & L. Tilling, eds., *The Correspondence of Isaac Newton* (Cambridge,1976), Vol.6, p. 193.
(2) Bancroft Brown, *Notes & Queries* (The Royal Astro. Soc. of Canada, 1927),Vol. 21, p.43-45.

(3) Lisa Jardine, *On a Grander Scale* (London, 2002), p.557.
(4) Ibid., pp.460-1.
(5) Aaron Klug, *Notes* (Royal Society, 1999), 54 (1) p.100.
(6) Isaac Newton (anon.), (R.S. Phil. Trans., 1715), Vol.29.
(7) E. Forbes, L. Murdin & F. Willmoth, eds., *The Correspondence of John Flamsteed* (Bristol, 1997), Vol. 3, p.740.
(8) Richard Westfall, *Never at Rest* (Cambridge, 1980), p.652.
(9) Francis Baily, *An Account of the Revd. John Flamsteed* (London, 1835/1966), pp.313, 316, 739-40.
(10) R. Hall & L. Tilling, eds. Vol. 6, p.212.

CHAPTER 27

Whilst the new South Sea Company took every opportunity to advance its cause, the Royal African Company found itself outmanoeuvred and fighting for its life. Court cases and internecine disputes had left it legally bankrupt and only by inflating the value of its fixed assets did it manage to keep its head above water.

According to *official* figures, a person subscribing to £100 (£20,000) worth of stock at inception in 1672 found his investment had shrunk to £60 by 1713. Over the years the investor would also have been called on to contribute £540 to support the company but would have received dividends of only about £215, so the RAC had over the years actually lost him £365 (£73,000). In the year 1711, the RAC had somehow managed to transport fewer slaves across the Atlantic than *Falconberg* alone had done in one trip seven years earlier. This at a time when the company's fleet probably consisted of more than 40 under-employed ocean-going vessels.

Meanwhile the SSC had acquired the monopoly on Britain's newly established rights to sell slaves to Spanish colonies in South America and the West Indies by underwriting almost £10 million (nearly £2 billion) of government short-term war debts. The company raised this enormous sum by selling more SSC stock. This was an attractive proposition to potential investors because the Bank of England would guarantee to pay over 5% per annum interest on the debt, which is why the business of actually trading in slaves became of secondary importance for a time.

The SSC had no holding forts in West Africa and no vessels in which to transport slaves because the directors had assumed the RAC could do all that for them, and they did, initially. However, the profit margins were too small to allow for the proper upkeep of their holding forts and the SSC was soon dealing with anyone willing to fill their order book.

But if persons had taken the time to study the history of slave trading with Spain, they would have discovered that it had always been unprofitable. Needing some high profile publicity to promote their stock further, King George I was persuaded to become Governor of the SSC and two of his mistresses somehow became influential stockholders. History began to repeat itself once more.

Early in 1720, parliament voted to let the SSC take over a further large portion of the national debt, many members having been bribed with cheap or free stock worth £1.3 million. The SSC was authorised to issue even more £100 stock certificates to raise this huge sum and, importantly was permitted to set a higher conversion price for the takeover rate. With £100 shares now officially valued at £135, £20 million of government debt could be converted by issuing less than £15 million of new stock, leaving over £5 million for extra fund raising, paying dividends, bribery and pocketing. The bubble was being inflated, but the company was still solvent because of the Bank of England's contracted annual interest payments and the inflated official price of the stock certificates.

Everyone wanted a slice of the action in much the same manner of the 'dot.com' boom of the 1990's. £100 stock certificates were soon being traded in Garraway's coffee house in Exchange Alley at £175 and bankers were advancing loans to investors to enable them to purchase even more stock. But then others got in on the scam, offering to sell a clearly gullible public shares in all manner of dubious schemes.

Parliament hastily passed a law requiring all joint stock companies to possess a Royal Charter and the SSC, which held one of course, had neatly disposed of the opposition who were diverting the funds of potential investors. Its stock values began to climb again, until by July 1720 the £1,000 per £100 certificate barrier had been breached.

But now the shrewd investors began taking their profits. The South Sea bubble began to deflate as speculators who had bought stock on credit defaulted on loans; slowly at first but then it burst as bubbles inevitably do. The first example of stock manipulation on a grand scale was followed by the first and one of most spectacular stock market crashes of all time.

Originally built entirely on the back of eagerly anticipated profits from the trans-Atlantic slave trade and the assumption that the longitude at sea problem would be solved now such an enormous prize was on offer. By the end of 1720 the stock value had returned to par and in 1723 the Royal Society still held £900 (£180,000) worth of SSC bonds [1].

With both the SSC and the old RAC now effectively cut down to size, the private shippers who had been waiting eagerly for their chance, quickly filled the void and took effective permanent commercial control of the British slave trade, the sea ports of Bristol and Liverpool being the main beneficiaries.

Parliament, the Bank of England, Britain and even the SSC survived the scandal although many members of parliament were forced to resign. Having lost many battles whilst trying to win more than its share of the international slave trade, Britain somehow won the overall war and soon became the largest slave-trading nation on Earth - all achieved without being able to determine a ship's position on the high seas with any degree of accuracy. And still the motions of the Moon remained unpredictable, but only just.

Shortly before the spectacular crash, the uncompromising rheumatoid irascible first Astronomer Royal had died at the age of 73 and was buried at Burstow where he had been rector for 35 years. At last the position of Astronomical Observer to the King had fallen vacant. Edmond Halley was the obvious choice and King George thought so too. If anyone could solve the motions of the Moon and in so doing also properly solve the longitude at sea problem and thus make him a profit, Halley was the man, longitude prize or no.

Edmond Halley's comments, published twelve years after his appointment to the post of Astronomer Royal are noteworthy. *'On Mr Flamsteed's Decease, about the Beginning of the Year 1720, his late Majesty King George I. was graciously pleased to bestow upon me the agreeable Post of his Astronomical Observer, expressly commanding me to apply myself with the utmost Care and Diligence to the rectifying the Tables of the Motions of the Heavens, and the places of the Fix'd Stars, in order to find out the so much desired Longitude at Sea, for the perfecting the Art of Navigation.'* [2]

The second Astronomer Royal, now a double *ex officio* commissioner of the Board of Longitude, had been commanded by his liege lord, not to explore the mysteries of the universe but to solve the longitude at sea problem using the same wording as had been used 45 years earlier (chapter 6), and for the same reasons. Despite the establishment of the massive public longitude prize, Halley had also been instructed to solve the problem (on a £100 annual salary before tax) in order to ensure the profitability of a private company of which King George I was governor.

However, because Flamsteed's widow had removed the observatory's instruments, steadfastly maintaining they had been the personal property of her late husband, Prince George having defaulted on his compensation undertaking, Halley too started work with limited resources. *'....... here I might have thought myself in a Condition to put in Execution my long projected Design of compleating my Abacus, or Table of the Defects of our Lunar Numbers; but on taking Possession, I found the Observatory wholly unprovided of Instruments, and indeed of every thing else that was moveable, which postponed my Endeavours until such Time as I could furnish myself with an Apparatus capable of the Exactness requisit. And this was the more grievous to me, on account of my advanced Age, being then in my sixty-fourth Year, which put me past all Hopes of ever living to see a compleat period of eighteen Years Observation.'* [2]

Flamsteed, in death, had at last succeeded in getting the better of at least one of his enemies and Halley was only to manage serious lunar studies for a half cycle of nine years (some 1,500 observations nevertheless) before infirmity took its inevitable toll.

Although Halley had previously obtained access to most of Flamsteed's data whilst preparing the stubborn man's material for publication, he was now at last able to properly check on their veracity after impatiently waiting for more than four decades. Immediately he discovered the gaps in Flamsteed's lunar data; the angry old man had registered none at all in 1716 for example.

Halley also discovered Flamsteed's secret; that the north-south wall, which Hooke had built, had been slowly sinking at the northern end for years; which called for a completely new stone wall to be constructed. His new appointment was to be no sinecure.

(1) Henry Lyons, *The Royal Society 1660-1940* (Cambridge,1944), p.128.
(2) Edmond Halley, *A Proposal ...for Finding the Longitude at Sea...*(R.S. Phil. Trans.,1731),Vol. 37, p.185-195.

CHAPTER 28

Because Flamsteed's widow had removed the big mural arc along with every other instrument in Greenwich Observatory including the pendulum clocks, Halley's first priority was to obtain his own 'state of the art' mural quadrant and to set it up on a new wall. Apparently he could have purchased Flamsteed's instrument from Margaret Flamsteed but wisely decided against this.

Halley's new 8ft. radius instrument (nowadays fixed on the wrong side of the wall) was manufactured by Jonathan Sisson and his friend George Graham divided the scale on the limb to greater accuracy than ever before. The instrument cost £204 and was paid for out of a £500 grant from Board of Ordnance, which took five years to make up its mind! So unfortunately this instrument was not fully functional until 1725. In truth the Board of Longitude should have supplied the grant, the instrument being essential to the quest.

As the new Astronomer Royal at last began to make his presence felt, Margaret Flamsteed succeeded in publishing most of her late husband's life's work, having had to get some of the all- important plates for the star maps engraved in Holland. It had not been easy but she had received some help from Whiston and a great deal from Flamsteed's former assistants Joseph Crosthwaite and Abraham Sharp.

Following the forced 'official' publication of *Historia coelestis* in 1712, Flamsteed had quietly continued collating the rest of his material and had by the time of his death already prepared two of his three intended volumes for the press, having also managed to repossess 300 of the 400 'spurious' undistributed copies of *Historia coelestis* in 1715.

Flamsteed's authorised version of his full star catalogue, the three volume *Historia coelestis Britannica* was completed by 1725, 47 years after intimating to his patron Moore that matters were well in hand, and six years after his death; sadly the volumes still contained errors. But the fruits of Flamsteed's secret mission, *Atlas coelestis* which contained no less than 26 massive 24x20 inch plates of constellations plus two of planispheres, was not published until around the time of the death of his wife in the summer of 1730.

Her will left not so much as a copper farthing to Crosthwaite or Sharp to compensate them for the years they had worked for her without charge [1].

Meanwhile Newton too had died - sadly in agony from a stone in the bladder - on the last day of March 1727; still owning a large amount of stock in the South Sea Company. His body was laid in State in Westminster Abbey before being interred in a prominent position in the nave, his theory of universal gravitation still unconfirmed. Scientists of all nationalities and not a few coin counterfeiters breathed a huge collective but muted sigh of relief.

In his final years, Newton had discovered that the future King George II and his wife Caroline were interested in chronology and numerology. Encouraged by Caroline he had gone so far as to prepare a manuscript which linked the timing of historical/mythical events such as the Argonauts expedition and the sacking of Troy via changing stellar positions caused by the Earth's precession, concluding that the former occurred in 934 B.C. A French pirated transcript later caused considerable furore and Halley was called on by the Royal Society to defend Newton's astronomical evidence, which he did very convincingly in his usual efficient style [2].

A year after Newton's death John Conduitt had the manuscript *(The Chronology of Ancient Kingdoms Amended)* published and five years later Newton's rambling *'Observations upon the Prophesies of Daniel'* also appeared in print.

In those circumstances it is hardly surprising that from the moment Newton's ornate monument was erected in Westminster Abbey in 1728, people have been carefully examining this for hidden messages presumably hoping to finding further biblical clues which might suggest a date for the second coming of Christ or more depressingly, Armageddon.

Whilst Halley's professional opinion of Newton's scientific achievements never appear to have wavered he must have experienced some growing resentment ever since the first Savilian appointment issue in 1691 when Newton, who clearly owed him an enormous debt of gratitude over the publication of *Principia*, had badly let him down by supporting Gregory. This and any number of disputes that Newton had involved him in had probably done irreparable damage to their personal relationship.

Back in 1686 Halley had written a Latin ode to Newton in the preface to the first edition of *'Principia'* [3].

John Conduitt was one of the many to heap praises on Halley's verse construction and when Richard Bentley took it upon himself to make alterations to this ode in the second edition, Halley was not amused and according to Conduitt, Newton was also put out by this [4].

The wealthy Conduitt had commissioned Newton's monument and had a hand in composing the epitaph. Why not ask Halley for advice on this?

However tempted, Halley could hardly include a hidden message that could be deciphered readily, his own well-earned reputation was far too precious.

The only method he could use was one that, if deciphered could be attributed to mere chance, just as Kepler's erroneous decipher of that Galileo anagram had been and just as Newton's own pseudonym could have been. If Halley *was* privy to the method used on the Shovell monument, how could he embed a signed mildly disparaging remark on Newton's own monument?

If he was tempted, as a mathematician and defender of *some* of Newton's numerology claims, surely he would feel obliged to use a key based on that self-same numerology.

Nothing too disrespectful; a gentle but final retaliation for all those years of being taken for granted perhaps? 'IN' was relatively easy but 'EH' was much more difficult.

The epitaph on Newton's monument was to be in classical Latin and there was little opportunity to use capitalization inappropriately. However, it can be demonstrated that there is a hidden message within the wording on Newton's monument based on a numerology key that takes all these points into account. This is the full text and layout of the epitaph:-

H. S. E.
ISAACVS NEWTON Eques Auratas .
Qui anami vi prope divinâ .
Planetarum Motus, Figuras ,
Cometarum femitas, Oceanique Æftus ,
Suâ Mathefi facem praeferente .
Primus demonftravit :
Radiorum Lucis diffimilitudines ,
Colorumque inde nafcentium propietates ,
Quas nemo antea vei fufpicatus erat, perveftigavit .
Naturæ, Antiquitatis, S. Scripturæ ,
Sedulus, fagax, fidus Interpres
Dei O.M. Mageftatem Philofophiâ afferuit ,
Evangelij Simplicitatem Moribus expreffit .
Sibi gratulentur Mortales ,
Tale tantumque exftitiffe
HVMANI GENERIS DECVS.

NAT. XXV. DEC. A.D. MDCXLII. OBIT. XX. MAR. MDCXXVI.

H. S. E. is an abbreviation for "Hic Sepultus Est" (Here lies buried). The total of numbers in the text, all relating to Newton's dates of birth and death (2+5 1+6+4+2, 2+0 1+7+2+6) add up to 38. Although Newton died in 1727, his year of death is given as being 1726 according to the civilian calendar (see Notes on dates).

Ignoring only the Roman numerals and counting every word or abbreviation in reverse from MAR (1) through to H (72) one can arrive at an apposite gentlemanly message which finishes with an identity. This is the only way Halley (or someone on his behalf) could possibly have included such an awkward set of initials in a Latin text. *Generis Humani gratulentur. E.H.* - Rejoice the human race. E.H. (8,9,1+4,7+0,7+2=38). Taking individual Latin words out of context is clearly not very satisfactory but then neither was Galileo's second anagram.

Newton's ornate edifice in Westminster Abbey includes carvings depicting all his scientific achievements bar one; his marine angle-measuring device.

(1) Francis Baily, *An Account of the Revd. John Flamsteed* (London, 1835/1966), p.363.
(2) Edmond Halley, *Remarks upon Some Dissertations...* (R.S. Phil.Trans., 1727/1728), Vol. 34, pp.205-210, Vol. 35, pp.296-300.
(3) Alan Cook, *Edmond Halley* (Oxford, 1998), pp.442-3.
(4) Ibid., p.165.

CHAPTER 29

With his magnificent new mural quadrant up and running, Halley invited Queen Caroline, wife of George II (George I had died four months after Newton in 1727) to visit the Greenwich Observatory in 1729. Having been informed that the Astronomical Observer's annual salary was a paltry £100, she was dissuaded from having it increased by Halley himself who, so the story goes, suggested that in future others might seek the position for gain rather than for the love of knowledge if a higher salary were on offer. A great pity no one had suggested increasing Flamsteed's salary instead of forcing him to take on part-time employment. Or had his low salary and lack of equipment caused in part because of his intransigent nature, suited the purpose of his hidden agenda?

Queen Caroline solved the problem of how to reward Halley by getting the Navy to do it for her. He was still a post-captain, technically on the Admiralty payroll, but had, like many in this position, never received his entitlement. This was speedily rectified by granting him a naval pension [1] and Halley was now a life-long employee of both the Admiralty and the Board of Ordnance and a Board of Longitude commissioner. An indebted member of the armed forces and a civil servant, he would never be able to comment on the Shovell disaster, the Longitude Board deception or Newton's marine angle-measuring instrument.

Now a series of what can best be termed embarrassing unhappy coincidences befell Halley. Whatever the manner of construction, Newton's device was clearly capable of accurately measuring angles as wide as 90° to one minute of arc from the rolling deck of a ship on the high seas. Equally clearly, it had remained a State secret since August 1699. However, this particular secret was always going to be difficult to contain, possibly because the instrument had been demonstrated at a Royal Society meeting, but certainly because it had been frequently used aboard *Paramore*. Halley had taken sights in the presence of the crew, which had unfortunately included Edward Harrison, an expert navigator, now dismissed the service long since but quite capable of behaving vindictively if still alive and out of range. The instrument may also have been inspected in Brazil by Hardwick, the Royal African Company agent.

Early in May 1731 and in the process of compiling a Greenwich Observatory progress report for the RS, Halley was informed that John

Hadley had invented a new type of angle-measuring device and was intending to lay claim to a portion of the Queen Anne longitude prize. How much his informant revealed about the instrument is not known but Halley could easily have guessed the twin-mirror double reflection principle would be involved. Halley now found himself in an extremely embarrassing position, made worse because John Hadley was not only also a vice-president of the society but was at the time serving with him on the RS Council.

The manner in which Halley handled the Hadley affair without lying, without ruining the reputation of Secretary to the Admiralty, Josiah Burchett MP, and without divulging State secrets was as masterful as the manner in which the late Prince George of Denmark had handled the Shovell disaster.

On the 17th May and before Hadley could announce details of his invention, Halley delivered a long paper to the RS which was based on an appendix he had previously written for the second edition of Street's *Caroline Tables* (chapter 25) and which had offered a lunar-based solution to the longitude at sea problem. Included was a damning indictment of his predecessor; one of the most outspoken and entirely justified condemnations of a fellow scientist ever made at a RS meeting; *'Mr.* Flamsteed *was long enough possessed of the* Royal Observatory *to have had a continued Series of observations for more than two* Periods *of eighteen Years; by which he had it in his Power to have done all that could be expected from Observation, towards discovering the* Law *of the* Lunar Motions. *But he contented himself with sparse Observations, leaving wide Gaps between, so as to omit frequently whole Months together; and in one Case the whole Year 1716.'*

This was followed by an overview of his own first 10 years of lunar observations as Astronomer Royal at Greenwich and Halley concluded with the following remark referring to John Hadley's angle-measuring device. *'It remains therefore to consider after what Manner Observations of the Moon may be made at Sea with the same Degree of Exactness: But I am informed from a very good Hand, that a most ingenious & worthy member of this Society has had Thought & taken some Pains about it, not without a Prospect of Success, which I heartily wish him. What I have meditated thereon I hope to speedily lay before you.'* Quite what Halley was meditating on he did not disclose.

Halley then carefully absented himself from the next meeting at which Hadley was due to make his announcement but arranged for his son Edmond, who was not a fellow, to attend in his place. If the instrument was based on Newton's double reflection principle Halley could hardly have remained silent if present.

John Hadley, although not demonstrating a working model, verbally communicated the description of his new invention; details of a twin-mirrored angle-measuring device, which Hadley stated, could be used to solve the longitude problem by the lunar distance method in the very near future thanks to Halley's Greenwich research. Hadley's appreciation did nothing to simplify matters for Halley.

Two weeks later the Astronomer Royal, having searched his conscience (and almost certainly conferred with Burchett) interrupted a reading of the minute regarding Hadley's announcement, stating that *'Sir Isaac Newton had formerly Invented an Instrument for the same purpose founded on the same principle of a double Reflection, and that he* [Newton] *Communicated some account of it to the Society in the Year 1699.'* Halley could have produced an accurate drawing, advised the gathering how he had successfully used the instrument to determine latitude on the three *Paramore* voyages on more than 180 occasions or even mentioned that Newton had actually shown it to fellows - had he been permitted. As it was he had had plenty of time to check the records to confirm that the Journal book entry gave away no secrets and that the *'some account of it'* did not mention the principle of double reflection. Halley's statement formally registered Newton's prior claim yet told his RS colleagues nothing.

The RS secretary Cromwell Mortimer then searched the records (presumably having received no help from Halley) and discovered that Newton had indeed *'communicated some account of it'* on 26th August 1699 but advised fellows that there was *'nothing particularly Expressed concerning the Construction of it so as to give any Light or Direction to know what it was any more than in general some improvement of the Common Quadrant used at Sea.'*

Mortimer failed to mention that the Journal book entry (chapter 18) also stated that Halley had actually used Newton's instrument at sea. Halley remained silent.

On 7th June, Hadley demonstrated his invention (figure 9) to a gathering of fellows, who by all accounts were mightily impressed; again Halley absented himself and maintained his silence. The RS September publication of Hadley's paper *The Description of a new Instrument for taking Angles* [2] also included a drawing of a second Hadley instrument (figure 10). Because of Halley's unhelpful silence, Mortimer then conducted another trawl through the RS archives and arrived at the illogical conclusion that the 1699 Journal book entry which clearly stated that Newton *'had shewed a new instrument contrived by him* [Newton] ...... *being the old instrument mended of some faults'* was really a reference to Halley's 1692 instrument (figure 5, chapter 13).

Mortimer sent a copy of Halley's drawing to Hadley informing him (correctly) that he was doubtful whether Halley's instrument incorporated his double reflection principle. Halley was then pressured into make an 'on the record' statement to clear the air but Mortimer's mistake coupled with the primitive design of both Hadley marine octants had provided Halley with a neat way out.

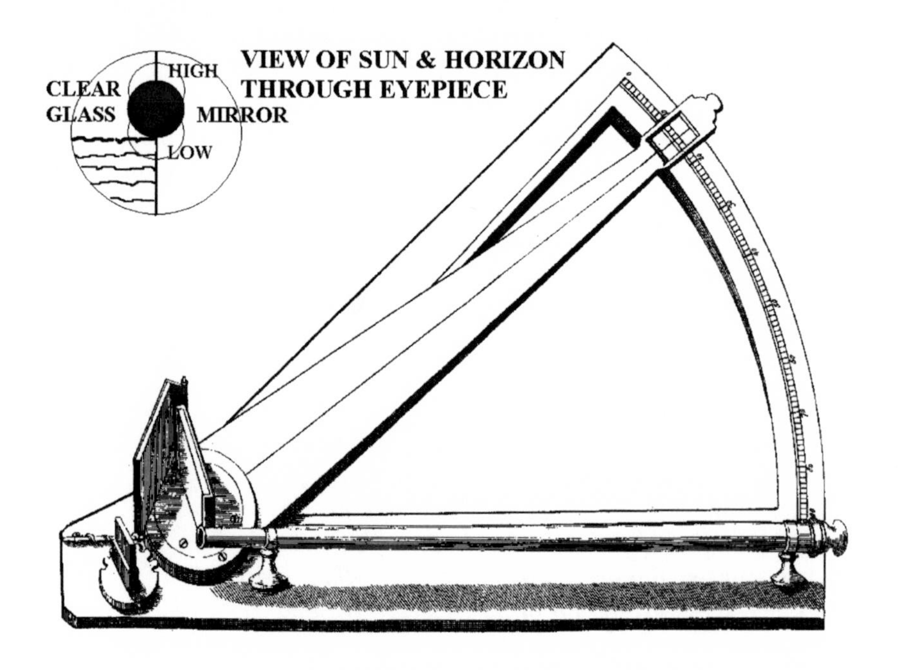

Figure 9. Hadley's basic marine octant.

This instrument could only measure angles to slightly better than ¼° of accuracy and was equipped with a sun filter. The fixed metal mirror had a piece of see-through glass attached to the outside edge. The view through the eyepiece would actually be inverted.

On 27th December 1731, the Astronomer Royal made a formal announcement at a RS meeting. '*Dr Halley took Occasion to say that he had considered the Construction of Mr. Hadleys new Invented Instrument for making Astronomical Observations on board a ship and was now well satisfied that it was much different from that which Sir Isaac Newton formerly Invented for that purpose & communicated to the Society.'* Everyone mistook this for a retraction of his earlier assertion that Newton was the inventor of the double reflection principle.

In fact all Halley had stated was that Hadley's instrument was *'much different'* as indeed it was, it was a poor imitation lacking a diagonal scale, made of wood and incapable of measuring angles to less than about 15 minutes of arc.

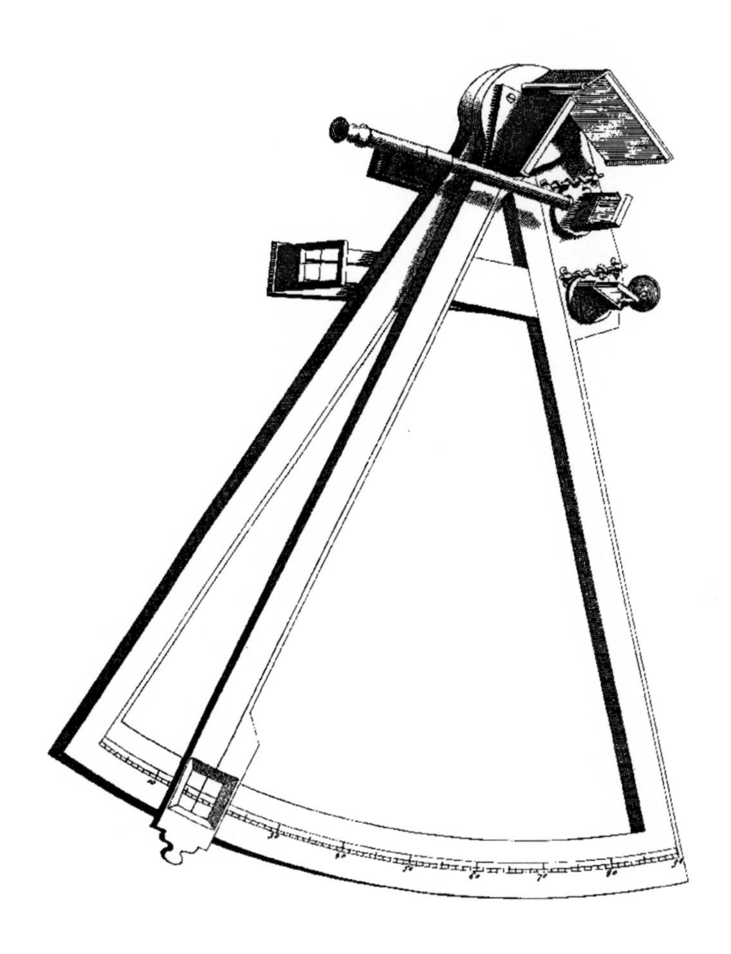

Figure 10. Hadley's second marine octant.

This instrument was fitted with a third metal mirror to allow the instrument to be used in the alternative back-staff manner. Both these mirrors had see-through glass attached to the outside edges. It was also fitted with a range-finding device. The mirror and telescope positions of this second device would become the standard design for future marine octants/sextants.

John Hadley and his supporters were further encouraged when Halley's Greenwich Observatory progress report was published shortly after his 'retraction' [3] and discovered Halley had changed the wording of his final sentence (see above for original version).
'It remains therefore to consider after what Manner Observations of the Moon may be made at Sea with the same Degree of Exactness: But since our worthy Vice President John Hadley, *Esq; (to whom we are highly obliged for his having perfected and brought into common Use the* Reflecting Telescope*) has been pleased to communicate his most ingenious Invention of an Instrument for taking the Angles with great Certainty by Reflection, (*Vide Transact. *No 420.) it is more than probable that the same may be applied to taking Angles at Sea with the desired Accuracy.'*

Again Halley had avoided the pitfalls. The second of Hadley's instruments (figure 10) incorporated innovative features (a third mirror for back-sights and a range-finding device) so Hadley was indeed the inventor of an ingenious instrument for taking angles by reflection. But there was still no endorsement of the double reflection principle. Embedded in the final paragraph is the second example of Halley's likely use of a cipher.

The number count is 4+2+0 = 6 and restricts any hidden message to a maximum of four words but provides an enormous 138 possibilities. However all but one have absolutely no relevance to the text. Yet *'No invention'* (a reverse count of capitalised words numbers 4 & 11) surely sums up Halley's private opinion of Hadley's instrument perfectly. In this case, there would have been no need to sign the message as the author's name headed the article. This he could privately reveal without having to specifically mention the Admiralty embargo should he be called on to explain to those with very long memories why he had failed to properly register Newton's prior claim.

In late August 1732, a letter dated 6th June addressed to my *'Esteemed Friend Dr. Edmond Halley'* was delivered to Greenwich Observatory via a ship's captain who had just completed a voyage from Philadelphia [4].

James Logan, a Philadelphian statesman and a well-respected mathematician had first met the Astronomer Royal in 1724 when visiting London. Now he was calling on that acquaintanceship to aid a friend Thomas Godfrey who, he claimed had invented a marine angle-measuring device based on the twin-mirror, double reflection principle.

Would Dr. Halley please submit this invention to the Board of Longitude for a possible award under the terms of the Queen Anne Act? Now the

BOL had two illegitimate claimants to deal with.

As far as is known, Halley failed to reply to Logan. If Godfrey's instrument should prove capable of accurate measurements, the Admiralty might find themselves committed to awarding development funds (or even a share of the prize) to an undeserving foreigner; either that or reveal the truth regarding Newton's invention.

So the Admiralty hastily laid on sea trials for the second Hadley model (figure 10). The illustrated instrument had only a primitive degree scale and Hadley must have added further sub-divisions prior to the trial. Nevertheless, although (Newton's) double reflection principle was sound, Halley knew Hadley's wooden instrument was still incapable of determining longitude by the lunar distance method where even a single minute of arc error can produce ½° of longitudinal error.

With the share of blame for the Shovell disaster now hanging only over Burchett's head (Haddock had died in 1714) and the embargo on Newton's one-off very expensive instrument still in force because of this ongoing threat to Burchett's considerable reputation, promoting Hadley's new cheap inaccurate wooden version over any likely foreign competition, was a way of reintroducing Newton's double reflection principle without either having to pay prize money or reveal past follies.

The test should have been designed to check whether Hadley's instrument was capable of taking accurate angular measurements from an unstable platform. If Hadley's claim to a portion of the longitude prize was to be seriously examined, all that needed be done was to check his octant on board a Royal Navy frigate against simultaneous solar/lunar observations made ashore with a fixed observatory quadrant.

However, such a simple arrangement would not suit Hadley because he knew his wooden instrument was not capable of measurements accurate enough to determine longitude by the lunar distance method, and this flaw would have been immediately exposed by such a test. This straightforward scientific method probably did not suit the Admiralty either because the proposed Godfrey version with its brass plate divisions and revolution counter might possibly be good enough to pass.

So an elaborate set of tests were set up to ensure that Hadley's instrument could appear to pass with flying colours despite being unable to do so. This would officially establish Hadley's priority over Godfrey and neatly lay to rest the troublesome Newton ghost.

Chatham, a small unarmed Admiralty vessel set off out into the choppy waters of the Thames Estuary on the 10th September 1732. On board were three official testers - John and Henry Hadley and James Bradley (who was

Halley's assistant at Greenwich).

Also in the party were the third Hadley brother George, two RS observers, and the official BOL observer. None of the testers had any experience of taking angular measurements from a rolling deck of a ship with a back-staff or any other instrument. The official BOL observer had a wealth of experience, being none other than the unfortunate Edmond Halley.

A number of measurements of the Sun's altitude were taken at various times using an unreliable watch, which had not been rated, and a Hadley instrument whose division errors were suspect. Back in London, the Hadley brothers converted the raw data into a plausible set of results by dishonestly adding or subtracting unverified watch and instrument errors. In this manner, 63 of the 81 sights were found to be less than 2' of arc in error.

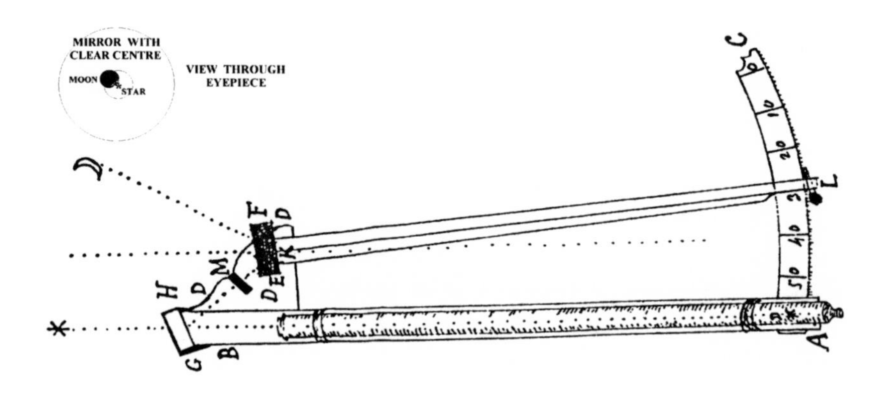

Figure 11. Godfrey's twin-mirrored angle-measuring device as drawn in Logan's letter.

Although this instrument appears to have been designed to measure maximum angles of 60° with an accuracy of about + or - ¼°, it could have been easily adapted to measure wider angles. A star is viewed directly through an un-silvered central section of the object mirror G-H fixed on the end of the brass base-plate D at an angle of about 70°. The image of the moon is reflected via a mirror E-F (attached to the moveable arm K-L) which is exactly parallel to G-H when that arm is set at 0°. Although not shown, Godfrey intended to engrave a diagonal scale on a brass insert along the limb A-C. He was of the opinion that this would have improved accuracy to within one minute of arc. The toothed ratchet screw he proposed (at L) would act as a revolution counter for fine measurements. M represents a glare filter. The wooden support frame A-B was to be about 40 or 45 inches in length. This instrument was an advance on the model tested in 1730 but had not been manufactured when Logan wrote to Halley in 1732 [4].

The only noon sight taken claimed to be accurate to *one second* of arc - 1/3600th of a degree! Needless to say, the official observer Halley did not certify the results and John Hadley did not receive a BOL award although the RS did publish the results [5], presumably because Halley had yet again declined to comment.

What of that overseas claim? Logan's protégé Godfrey was an amateur astronomer and glazier who, according to Logan had first tested his instrument in 1730. The drawing, which Logan had enclosed with his letter to Halley, is shown in figure 11; this was of a new prototype, not the tested instrument.

Godfrey's fixed mirror/glass included a small clear section through which a star was to be viewed. The instrument was designed to measure the angle between two heavenly bodies; the lunar distance method of determining longitude and hence the intended BOL claim.

Shortly after the *Chatham* trials and having received no reply from Halley, Godfrey and Logan submitted details to the RS of another Godfrey invention, the *'Mariners Bow'* (an improvement on the back-staff). They also mentioned Logan's letter to Halley, a letter that had been written to a BOL commissioner, not the RS.Nevertheless, this letter was read to fellows on 11th February 1734, some 18 months after Halley had received it.

If John Hadley had previously been unaware of the details, he now realised Godfrey had lodged a prior claim to 'his' invention. Instead of publishing the details, the RS simply published Godfrey's paper on his *'Mariners Bow'* invention [6].

His twin-mirrored device received scant mention in the nine-page document; no description and no diagram.

There is a further twist to this convoluted story. Godfrey had worked with George Steward, mate of the sloop *Trueman* on the construction and testing of his instrument which culminated in a test run to Jamaica in late 1730 [7]. There rumour has it, the instrument was shown to a relative of John Hadley but that evidence is circumstantial.

Be that as it may, there is a very good chance that the Royal Navy and or British Admiralty representatives in Jamaica would have got wind of a primitive twin-mirrored angle-measuring instrument being tested on board a New England vessel in 1730 a year *before* John Hadley suddenly managed to produce two instruments based on the same principle.

Whatever the truth, Halley had succeeded in avoiding the numerous pitfalls and revealed no State secrets and the Admiralty had sidestepped two claims on the longitude prize.

(1) Colin Ronan, *Edmond Halley; Genius in Eclipse* (London, 1970), p.238.
(2) John Hadley, *The Description of a new Instrument for taking Angles* (R.S. Phil. Trans., 1731), Vol. 37, pp.147-157.
(3) Edmond Halley, *A Proposal...for Finding the Longitude at Sea...*(R.S. Phil. Trans., 1731), Vol. 37, p.185-195.
(4) James Logan, *Letter to Edmond Halley* (R.S. private records, 1732), RS EL/L6/59.
(5) John Hadley, *An Account of Observations made on Board the Chatham-Yacht* (R.S. Phil. Trans., 1732), Vol. 37, pp.341-356.
(6) James Logan, *An Account of Mr. Thomas Godfrey's Improvement of Davis's Quadrant* (R.S. Phil.Trans., 1734),Vol. 38, pp.441-450.
(7) William Andrewes in *The Quest for Longitude* (Harvard, 1996), pp.401-2.

CHAPTER 30

Shortly after Mary, his wife of 55 years died in 1736, Halley suffered a minor stroke and lost the use of his right hand. His will made five months after his wife's death, left his estate to his son (who was to die in 1741) but left his books and papers to his two daughters whom he named as joint executrices. He also left instructions that he should be buried in the same grave as his wife.

The Astronomer Royal was sitting comfortably in his favourite chair in front of a warming fire on a cold January day in 1742, 100 years after the death of Galileo, the man who had discovered Jupiter's moons, which led to the solving of the basic longitude problem. A glass of wine was on a side table within easy reach and the clockmaker John Harrison's battles with the Longitude Board still in the distant future. However, Halley was, as Galileo had been, tired of the scientific infighting that had dogged his entire career.

His eyes closed and his head fell gently back. Was he dreaming the 'if only' dreams that haunt so many great men born before their time? If only the Greenwich Observatory had been properly planned and equipped in 1676 and if only Flamsteed had stuck to his allotted task. If only Newton had not used his own good nature and drawn him into unseemly disputes that had unfairly ruined reputations. Any number of reasonable little 'if's' which jointly might well have solved the longitude at sea problem in his lifetime.

Yet so many lives were still being lost at sea and all because there had not been enough rotten gunpowder. The Duchess of Portsmouth's yearly dress allowance could have built 20 observatories; or would it have been or 200..... or 2,000?... or was that the number of sailors drowned by that fool Shovell or the number of twin-mirrored angle-measuring instruments the Mint could have produced? His mind, so full of numbers all his life was at last perhaps having trouble remaining focused whilst still adding zeroes. 20,000.... 200,000... or was that the number of slaves lost at sea?

Let us hope his dreams did not drift into bitterness because this time he would not wake again; the finest astronomer of his age and the greatest oceanic navigator of all time had died in his sleep, whilst the Moon, unobserved for once, occulted two stars in the Virgo constellation. Neither was listed in Flamsteed's *Historia coelestis Britannica* or illustrated in *Atlas coelestis*, the work the stubborn man had devoted his life to producing on behalf of his Maker.

Cloudesley Shovell, Knight, Rear Admiral of England and Admiral and Commander in Chief of the Fleet had been buried in an ornate tomb in Westminster Abbey. Isaac Newton, Knight, Master of the Mint and President of the Royal Society had also been buried in Westminster Abbey, as would be his own close friend George Graham. But Edmond Halley, FRS, LL.D, Astronomer Royal, born in an English republic, yet employed by no less than seven successive British monarchs was buried without ceremony and according to his instructions, beside his wife in St. Margaret's Churchyard, Lee, a mile to the south of Greenwich Observatory. His bones would surely have felt uncomfortable lying anywhere near those of Shovell.

Following Halley's death, a hand-written note describing an angle-measuring device was somehow discovered amongst Halley's papers by RS vice-president Sir William Jones. Jones was a retired Royal Navy navigation instructor who, by another of those strange coincidences had been present at the battle of Vigo (chapter 22) before wisely deciding on a life ashore.

In 1709, Newton and Halley had written letters of recommendation in support of his effort to obtain the master-ship of Christ's Hospital mathematical school. He was unsuccessful mainly because of Flamsteed's opposition to anyone supported by either man and Flamsteed's assistant James Hodgson was awarded the post. Jones had later played a major role in supporting Newton in his unsavoury attack on Leibniz (chapter 26). A man who would certainly recognise the description of a marine twin-mirrored angle-measuring device and the great man's handwriting the moment he read Newton's note. He would also have realised the note's implications for John Hadley.

Jones was also an avid collector of Isaac Newton's mathematical papers but he never disclosed exactly how he had acquired this note, although he apparently claimed he took charge of it, thinking it belonged to the RS. This seems an unlikely tale because he then left the society with only a copy.

In early November 1742, immediately after the 75 year old Josiah Burchett retired from his Admiralty post Jones presented Newton's note [1] to RS fellows, none of whom had been present at that August meeting 43 years earlier at which Newton had demonstrated the actual instrument and had made that extraordinary longitude claim.

The undated note, together with an accompanying diagram had been given to Halley for comment in 1696 or possibly 1697 at the time he was Newton's deputy at Chester Mint.

By hindsight, this note laid out an idea for an instrument yet to be made in much the same way that Hadley and Godfrey would later describe and sketch instruments prior to manufacture. But in this case, the diagram was by now either lost or had been pocketed by Jones; sketches by Newton were rare items.

The full text [1] of Newton's note as published in the *Philosophical Transactions of the Royal Society, Vol.42, no. 465 pp 155-6, 1742* but with paragraph numbers added is set out in appendix 10.

The first paragraph read '*In the annexed Scheme, PQRS denotes a Plate of Brass, accurately divided in the Limb DQ, into 1/2 Degrees, 1/2 Minutes, and 1/12 Minutes, by a Diagonal Scale; and the 1/2 Degrees, and 1/2 Minutes, and the 1/12 Minutes, counted for Degrees, Minutes, and 1/6 Minutes*' ; the rest of the letter was equally technical.

This opening paragraph confirms Newton intended to make his instrument from metal, intended to make use of a diagonal scale that could measure angles to an accuracy of better than one minute of arc and most importantly intended to use reflection because of the statement that the half degree divisions would be doubled (by reflection) and count as full degrees.

The RS must surely have commissioned an expert familiar with twin-mirrored angle-measuring devices such as John or George Hadley to advise on a new interpretive diagram to accompany the publication. Certainly only such persons could have made sense of Newton's note. For example, how could anyone but an expert guess Newton was referring to a thin wedge of cheese shape when he wrote *'PQRS denotes a Plate of Brass'* or make any sense out of the diagonal scale division descriptions?

Unfortunately, although an expert obviously did advise the artist, by accident or design he made a confusingly poor job of it. Even worse, the drawing was then given to an engraver (figure 12) and the engraving was passed off as a copy of Newton's original (lost) *'annexed scheme'*.

The artist's wash drawing failed to match Newton's descriptive note on four major points (illustrated and discussed in appendix 10), the most obvious being that the assumed 45° arc was divided into 45 degree sections; no allowance was made for the reflection halving the space occupied by each degree division. This glaring mistake was corrected on the engraving but surely not by the engraver?

Despite the correction, the engraver then managed to create two new mistakes and the only area left partially uncorrupted was Newton's innovative twin-mirror double reflection invention. Indeed the illustration was so misleading that Newton's newly-discovered note now conveniently posed little risk to vice-president John Hadley's reputation, a reputation originally established 10 years earlier by the publication in the same journal of the seriously flawed *Chatham* tests (chapter 29).

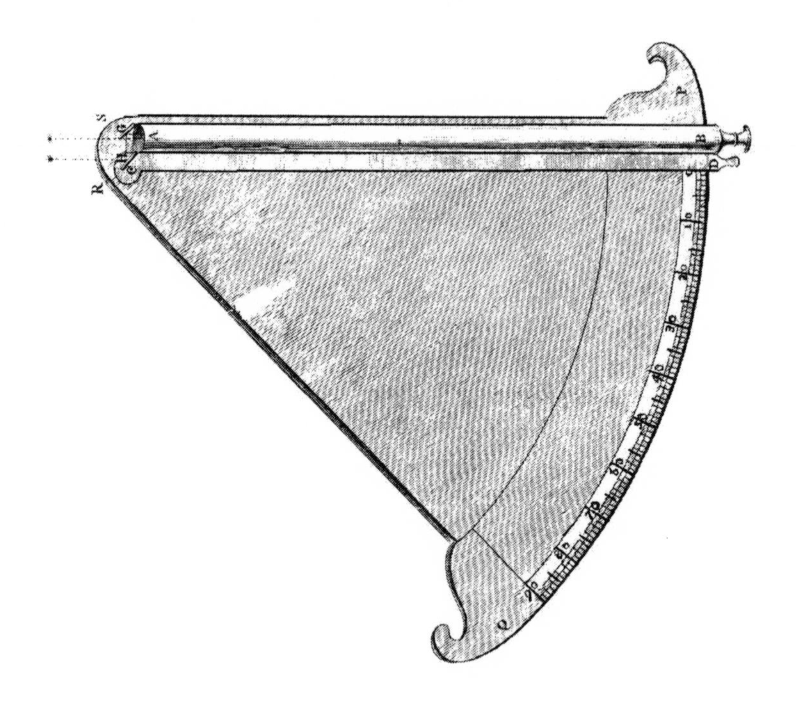

Figure 12. The Royal Society's steel engraving of Newton's marine octant. *The diagonal scale, carefully but confusingly described in the first and fourth paragraphs of Newton's letter was not included in either the artist's drawing or the RS engraving and this omission implied that accurate measurements were impossible.*

The zero end of the scale was shown as being nearest the telescope (see figure 13 for the correct setting) which meant that the instrument had to be used 'upside down' and the fixed mirror masked angles of more than about 45° making half the instrument useless. The two mirrors were not parallel when set at zero, the index pivot point was not in line with the edge of the index arm and the index mirror was not placed directly over that pivot point. The two curly end bits on the base plate (not mentioned in Newton's description) presumably intended to hang this cumbersome instrument from a ship's rigging, were placed in such a way as to point the large telescope towards the seabed. The 'three or four foot' telescope which Newton had suggested but who's length was of little importance, was assumed to be incorporated within the frame of the instrument, implying that the frame was at least three feet long, the solid brass instrument thus weighing about 30 kilograms. An instrument any expert would dismiss as primitive at best.

Despite the convenient 'doctoring' of the Newton paper, John Hadley still faced a potential financial setback. In 1734, he had successfully applied for an English/Welsh patent on his invention [2] and marine quadrants had since been manufactured under licence by a number of leading instrument makers.

The actual wording of Hadley's patent was vague and no illustration of any device had been included, which in itself was strange. Indeed the patent, which would expire in 1748, had been granted on the strength of the double reflection principle (appendix 11) rather than on a specific design; Newton's principle.

The patent also stated that if anyone were to provide proof that John Hadley was *not* the inventor the *'Letters Patent shall forthwith cease'.* GB patent number 550 was not annulled; presumably the instrument-makers paying Hadley royalties did not subscribe to RS publications.

One interesting comment made in the body of the patent, and which may explain why there was no protest, was that Hadley's invention *'May be of signal use to our Royal Navy and to the trade and navigation of our kingdoms.'* [2]

All of which somehow let John Hadley off the hook and consigned Newton's wonderful invention to the rubbish bin. In such circumstances, it is little wonder that no one was motivated to investigate that 1699 longitude claim or to examine the *real* instrument's effectiveness in determining *latitude* as carefully detailed in Halley's *Paramore* journals.

Halley had used Newton's marine octant very successfully during all three *Paramore* expeditions, repeatedly and accurately measuring angles as wide as 87°. Newton's published note, which placed the fixed mirror (G in figure 12) inclined 45° to the Axis of the Telescope, was not describing the instrument he loaned to Halley.

It almost certainly was not describing a working instrument at all. Halley, with his specialist knowledge of observatory instruments would have realised Newton's proposal was impractical for the purposes he intended the moment he read the note.

There were several design flaws, which must have been dealt with before Halley could have used Newton's device to successfully measure angles on the high seas in the very accurate way that he did. The mirrors had to be moved to the positions and angles shown in figure 13, otherwise wide angles could not be measured. Filters would have to be fitted to prevent the user from suffering retinal damage when observing the Sun, and the heavy index arm might have required the fitting of an improved locking mechanism. Certainly Halley would have considered adding clear glass to the outside of the fixed mirror and would have made scant use of a long telescope as he would immediately have discovered that the instrument worked well at sea with a very small version.

A few cross-staffs and back-staffs (including Godfrey's *'Mariner's Bow'*) had from time to time incorporated the diagonal method of division but the difficulty of marking out very fine lines on boxwood or ivory by hand was insurmountable.

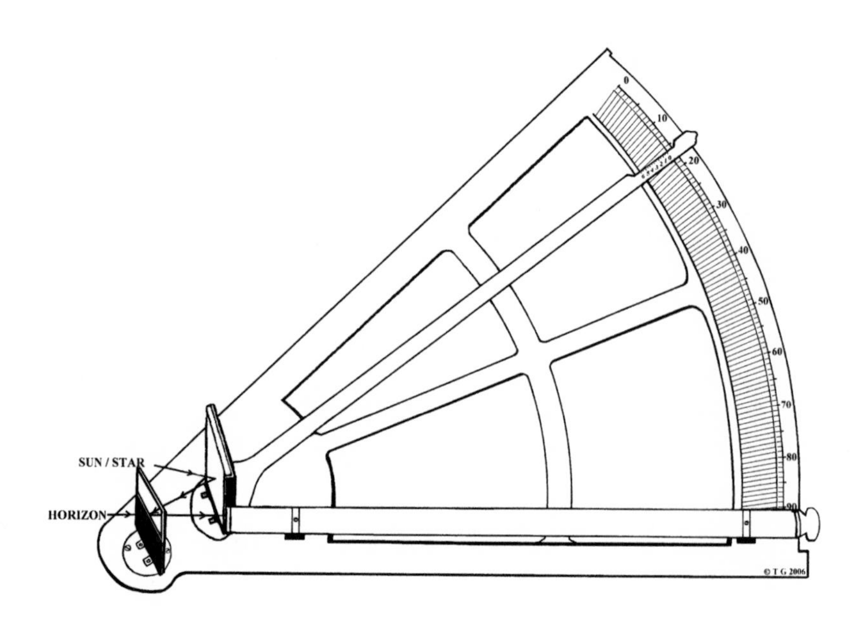

Figure 13. A drawing of the author's working model of Newton's brass marine octant.

The view through the eyepiece is as shown in figure 9, chapter 29 and the diagonal scale as detailed in figure 4, chapter 11. The glare filter between the two mirrors is omitted as is the index arm locking device and a handgrip. This version has a clear piece of glass fitted to the outside of the object mirror (instead of clear air), as in the Halley and Hadley designs.

This is the author's second attempt - the first incorporated the full-length 360 section diagonal scale as described in the first paragraph of Newton's note (see appendix 10 for discussion). The indented section of the index arm is divided into 60, each segment representing one minute of arc. The angle represented is 15°55'.

This problem, coupled with the fact that the tapering area occupied by the diagonals could not be perfectly bisected by a straight line, had always limited their effectiveness. Newton's access to the Mint engravers skills and to the foundry metalworking facilities (chapter 15) coupled with his own mathematical abilities, had removed that constriction. He must surely have cut straight diagonals (as in figure 13) and then either constructed a deviation table, or graduated the (60) marks on the index arm.

Without doubt Newton's basic instrument was his own invention and the innovative use of the two mirrors changed the primitive and unwieldy back-staff and the flimsy instruments of Hooke and Halley into a forward viewing, self-correcting, easy to use work of art that owed nothing to Hooke or Wren. As if someone had stripped down a Penny-farthing bicycle, analysed all the defects and then had produced a modern machine, complete with pneumatic tyres, chain drive, multiple gearing and cable-operated brakes.

How much Newton owed to Halley's practical experience (figure 5, chapter 13) and advice on the diagonal scale and mirror problems is impossible to determine. Most astounding was the fact that Newton, of all people, should have produced the instrument that would spur the quest for a complete understanding of the Moon's orbit (in order to determine longitude at sea by the lunar distance method). This device was the key to properly developing his theory of universal gravitation and he had first designed it and then had it manufactured in secret with only Halley to advise him.

When conducting the daily chore of determining *latitude* with the new John Hadley instruments, mariners discovered the two attractive features of the Newtonian design, which is why these eventually replaced the back-staff once Hadley's manufacturers had solved the scale division problems.

First, the Hadley octant could be automatically checked for errors by viewing a single object with the index arm set at zero (paragraph 5 of Newton's note - appendix 10). The two images of the one object (normally the horizon) should coincide; if not, because the mirrors have been knocked out of alignment and are no longer exactly parallel for example (as was claimed had occurred during the *Chatham* tests), make allowances (index error) or adjust a mirror until they do coincide. Exactly as is done with a modern marine sextant.

Second, rotating this instrument by several degrees did not disturb the coincidence of the two images; the superimposed image of the Sun's lower limb could be 'rocked' to kiss the horizon and as long as it just touched sometime during each roll, the measurement was accurate (paragraph 6 of Newton's note - appendix 10).

Commercial marine twin-mirrored angle-measuring devices never equalled the sophistication of Newton's sidelined instrument prior to 1758. In that year the instrument maker John Bird expanded the octant's angular measuring range from 90° to 120° and the marine *sextant* became the standard maritime angle-measuring device. He also eventually mastered the Royal Mint's secret of how to produce flat brass sheets free of impurities.

John Bird was also recognised as the first person to add a vernier scale capable of measuring to better than 1' of arc and in 1767 was awarded £560 (over £100,000) by the Board of Longitude in order to reveal his secrets and to teach English apprentices his special techniques for accurate scale division of astronomical instruments by hand [3].

By hindsight, this was a waste of money because less than seven years later Jesse Ramsden was awarded £300 for his dividing engine invention and a further £315 for the machine itself, which the commissioners then lent back to him. This machine enabled the degree scale to be cut quickly and accurately and brought the cost of a marine sextant within the reach of the pockets of most navigators *the Admiralty approved of.* Had Hooke been alive he would certainly have claimed prior invention on Ramsden's machine because he and Tompion between them had invented a similar dividing engine for cutting clock and watch cogs over 100 years earlier.

John Flamsteed and Isaac Newton had both died, missions unaccomplished. In truth, neither had failed their Maker and those that followed only succeeded in completing these tasks because of the groundwork laid by these two very different characters.

Flamsteed's star maps were completed relatively quickly but have only been quietly acknowledged and he is actually better known for the 'Flamsteed' glass invented by Hooke and the 'Flamsteed' star numbering system invented by Halley.

On the other hand, Newton's mission, which was completed in large measure by the efforts of the disciples who survived him, has been glorified. Yet his one and only wonderful invention - his twin-mirrored self-correcting marine octant - had been sidelined by the dubious interests of national security. And in so doing had completely eliminated Halley's important role in that instrument's development and delayed the introduction of a life-saving instrument by more than three decades.

Despite his self-effacing nature and his obscure burial site in a little-visited graveyard, Edmond Halley has always been remembered. Each succeeding generation is reminded of his name by the regular spectacular reappearance of 'his' comet, even if they have no idea that he was actually a scientist whose intellect outshone the brilliance of any comet. Above all else, Edmond Halley should be remembered for being *one of the most respected scientists* in an age of scientific greats.

Colin Ronan, an acknowledged expert on Edmond Halley wrote a classic assessment of the man in *'Edmond Halley; Genius in Eclipse'*. Some of his comments are apposite. *'It has been Halley's misfortune to be eclipsed by a man whose work he himself was mainly responsible for publishing. For his very ability to grasp the profound significance of Newton's research entailed a mathematical and astronomical knowledge and insight possessed by few men of any age... he performed this service for science...* '[4]

In the event that the reader has become confused by all the necessary technicalities, Warden of the Royal Mint Isaac Newton invented and constructed the marine twin-mirrored, self correcting angle-measuring device based on the double reflection principle, the single most important invention in the field of marine navigation in 3,000 years. Not RS vice-president John Hadley or the Philadelphian Thomas Godfrey, but ISAAC NEWTON.

Edmond Halley then used Newton's invention to accurately determine the latitude of the Scilly Isles, information that should have averted the tragic Shovell disaster of 1707. Halley also used Newton's instrument to accurately determine his position on the high seas, the first man ever to do so; not a mythical Chinese Admiral in 1421, or Christopher Columbus in 1492 or Amerigo Vespucci in 1499 or William Dampier or James Cook or John Harrison but EDMOND HALLEY - on 21st April 1700. End of story - almost.

(1) Isaac Newton, *A true Copy of a Paper found, in the Hand Writing of Sir Isaac Newton..* (R.S. Phil. Trans., 1742), Vol. 42, pp. 155-6.
(2) John Hadley, *Hadley's Patent* (The Patent Office GB 550, 1734).
(3) John Bird, *The Method of Dividing Astronomical Instruments* (London, 1767).
(4) Colin Ronan, *Edmond Halley; Genius in Eclipse* (London, 1970), p. 215.

EPILOGUE

Although Edmond Halley's death brought the story to a close, several mysteries remained. For example, what happened to Newton's marine octant and did Halley leave any other embarrassing legacies?

Halley, despite his lifelong diplomatic reluctance to put pen to paper, had clearly faced a dilemma in his final years. What was to become of the confidential material in his possession? For example, although he had obviously submitted accounts of his three voyages to the Admiralty, he would surely have retained copies. Destroy the sensitive and embargoed material perhaps? But this might mean that major scientific research conducted during some of his most productive years would be lost forever, those three *Paramore* journals in particular; the Admiralty were unlikely to publish details that would highlight their own incompetence.

His wife had died in 1736 and his naval surgeon son early in 1741. His eldest daughter Margaret had never married and his only other child Katherine was now a widow without issue and past child-bearing age, her husband having died in 1740. Therefore, the responsibility for protecting any unpublished data would fall on his two daughters who, under the terms of his will would jointly inherit his books and papers.

However, neither woman was equipped to hold her own in the murky world of scientific intrigue and Admiralty secrets. In the event it seems likely Margaret and Katherine were instructed to hold onto his unpublished data and only to pass this on if and when they thought the time right or profitable. Dampier - who had died in 1715 and was already being hailed as a great oceanic explorer - had profited from the publication of his journals; perhaps Margaret and Katherine could profit from their father's?

Margaret died a year after her father, and Katherine later handed a number of documents, including lunar data and trajectory calculations and copies of at least parts of Halley's precious journals to Lord Aberdour (the Earl of Morton and president of the Royal Society) shortly before her own death in 1765 hoping for a financial reward from the Board of Longitude commissioners, who took possession of them, published nothing but had the good grace to award her £100. Alexander Dalrymple then borrowed the 'journals' of Halley's first two voyages from the BOL and published the details in 1773 under the title *'Two voyages made in 1698, 1699 and 1700, by Dr. Edmond Halley, published from the original manuscript in the possession of the Board of Longitude.'* Dalrymple failed to return these loose sheets of paper, which were not in Halley's handwriting, claiming they had been lost in a fire when his house burned down. Subsequent events suggest this was indeed the case.

Over 100 years later in 1877 *different* copies of these first two voyages, which contained additional information and included Halley's signature, but were again not by his own hand, turned up at a Southeby's auction. However, included in the offering, which came from the estate of the late Sir Henry Weysford Charles Plantagenet Rawdon-Hastings, 5th Earl of Moira and 4th Marquess of Hastings, was a neatly presented original autographed account of Halley's 3rd voyage, the English Channel survey. Fortunately, the lot was purchased for the nation by the British Museum and later transferred to the British Library, but no one can account for how these came to be in the possession of the Rawdon-Hastings family. They were presumably either the Admiralty versions and/or yet other copies. Other material relating to Halley's RS publications had been retained by society Secretary John Machin during Halley's lifetime and these later somehow came into the possession of the Royal Astronomical Society. This material was stolen in the 19th century.

In 1981, The Hakluyt Society published the British Library material and Halley's English Channel survey saw the light of day for the first time, 280 years after it had been written. Edited by Norman J W Thrower and published under the title *'The Three Voyages of Edmond Halley in the Paramore 1698-1701',* this enthralling book was republished in 1999 and is still in print.

So although the basic details of Newton's invention had been in the public domain since 1742, the *full* story relating to Halley's successful usage of it did not become available for scrutiny until 1981 and even then attracted little attention owing to the lack of accompanying evidence and an inability to validate Halley's claims prior to the introduction of SkyMap© type computer software. However, Halley could never have imagined the invention of the computer, let alone a system that could hunt back in time hundreds or thousands of years and display all the details of the heavenly bodies as viewed by him at a specific moment in time in the middle of the southern Atlantic Ocean. Therefore, logic would suggest that the precious journals might not have been the only written evidence he left behind. Moreover, having lived in an era of vicious unethical scientific and governmental infighting and secrecy he would have known better than to trust all to the good fortune of his daughters.

But *if* Halley had contrived to place a mild admonition on Newton's magnificent edifice in Westminster Abbey, the odds of someone eventually deciphering it were very real, given that people were continually and minutely inspecting it for clues to the second coming of Christ. And when someone did uncover his signed note on that memorial they would surely examine his own memorial to check if a similar method of encryption had been employed there as well.

Either Halley had a hand in composing his own epitaph or his daughters had been told of their father's specific method of encryption because if one examines Halley's tombstone there are, rather confusingly, two examples of ciphers using that self-same numeric type of key.

SUB HOC MARMORE
PLACIDE REQUIESCIT CUM UXORE CARISSIMA
EDMUNDUS HALLEIUS, LL.D.
ASTRONOMORUM SUI SECULI FACILE PRINCEPS
UT VERO SCIAS LECTOR
QUALIS QUANTOSQUE VIR ILLE FUIT,
SCRIPTA EJUS MULTIFARIA LEGE;
QUIBUS UMNES FERE ARTES ET SCIENTIAS
ILLUSTRAVIT ORNAVIT AMPLIFICAVIT
ÆQUUM EST IGITUR
UT QUEM CIVES SUI VIVUM
TANTOPERE COLUERE.
MEMORIUM EJUS POSTERITAS
GRATA VENERETUR.
NATUS MDCLVI.
EST. A. C.
MORTUUS MDCCXLI II.
HOC SAXUM OPTIMIS PARENTIBUS
SACRARUNT DUÆ FILIÆ PIENTISSIMÆ
ANNO C. MDCCXLII.

Inscription on Halley's weatherworn marble memorial slab set into the wall of the Camera Obscura at Greenwich Observatory.

This English translation is set on a plaque to one side:
Beneath this gravestone, Edmond Halley, unquestionably the most eminent of the astronomers of his age, rests peacefully with his dearest wife. So that the reader may know what kind and how great a man (Halley) was, read his various writings in which he dignified, embellished and strengthened almost all the arts and sciences.
And, therefore, as he was a man so greatly cherished by his fellow-citizens during his lifetime, so let a grateful posterity venerate his memory. Born in the year of our Lord 1656. Died 1741/2. This stone was consecrated to excellent parents by two devoted daughters in the year 1742.

Beneath the Latin inscription are these which were added later [1].

Here is Alfo interr.^{d} M^{rs}. MARGARET HALLEY,
The Eldest Daughter of the Above D^{r}. HALLEY,
She died on the 13th of October 1743.
In the 55th. Year of her Age.

Also Mrs Catharine Price Youngest
Daughter of the above Dr Halley
who died Nov the 10th 1765 Aged 77 Ye^{s}.
and Mr Henry Price her Husband

The total number count of the original Latin inscription including the two hardly noticeable '11's' following the 1741 (Julian) date of death is 47. Reading in reverse, the method forced on Halley in order to include his initials in the Newton epitaph, nothing fits. Reading the right way as in the Shovell cipher produces:-

EDMUNDUS SCRIPTA EJUS MULTIFARIA LEGE; Protector/Guardian (from Anglo-Saxon) read his various writings. Taken in isolation this is not a very convincing hidden message given that the text makes a similar request in plain language, although it can be argued that the 'Protector/Guardian' emphasis turned a simple eulogy into a specific plea.

If Halley left a cryptic note on his tombstone requesting an historian to read his various writings, he would surely have also left a clue as to their whereabouts. Bearing in mind Halley had left Newton's prototype instrument notes where they could be discovered and published after his death, he would have expected nothing less than a re-appraisal by the RS of John Hadley's claim to have been the inventor. He was not to know there would be a cover-up. Expecting nothing less than a backlash over his unethical support of Hadley, he must have made provisions for revealing his *'no invention'* cipher. But of course, there was no backlash and the 420=6 cipher presumably remained undisclosed. Had it been revealed, someone might have applied it to Halley's tombstone. A forward number count of the first three words 'SUB HOC MARMORE' - under this stone - also = 6. Did this mean the writings he was asking to be read were buried with him? If so, perhaps that is where Newton's instrument was buried also?

The incised slab found its way to Greenwich Observatory when it was removed from the top of the family vault in St. Margaret's Churchyard, Lee, by the Lords Commissioners of the Admiralty in 1854 when the entire neglected tomb was renovated and a replica top installed.

The Admiralty again cleaned the vault in 1909, but now the replica stone slab is once more virtually unreadable. Covered in lichen and bird droppings with a piece broken off one corner, no casual observer would know it was the last resting place of the finest navigator of all time. For those interested, the tomb is in the old churchyard near the east wall across a busy road from the modern church. It can just be recognised by the badly eroded wording on the west end:

RESTORED
BY
THE LORDS COMMISSIONERS
OF THE ADMIRALTY.
1854

On the hidden east end is a memorial to John Pond (1767-1836) a later Astronomer Royal who was buried in Halley's tomb. Because of the sheltered position, this inscription is the only one that is still clearly legible.

After all the years of infighting, claims and counter-claims and official secrecy, the lunar orbit could still not be accurately predicted at the time of Halley's death. Regular determination of longitude on the high seas by astro-observation - or by any other method - was still not possible. The *average* mariner could no better determine his longitude in 1742 than he could have done in 1642. Latitude, thanks to the marine twin-mirrored octant for those navigators who could afford a Hadley instrument, yes; but longitude no.

Many readers, having read Dava Sobel's *'Longitude'* might think that this puzzle was eventually solved by John Harrison's marine chronometer in the late 1760's when, against all the odds and despite the appalling behaviour of the 5th Astronomer Royal, Neville Maskelyne, Harrison managed to 'win' the great longitude prize. Unfortunately, the truth is somewhat different. Harrison did *not* win the prize and it was actually Maskelyne who finally cracked the case with a little posthumous assistance from Newton and Halley. Sobel during her extensive research had not appreciated the underlying truth in Newton's 1721 letter to Josiah Burchett (chapter 26):-

'And I have told you oftener than once that it is not to be found by Clock-work alone. Clockwork may be subservient to Astronomy but without Astronomy the longitude is not to be found. Exact instruments for keeping time can be usefull only for keeping the Longitude while you have it. If it be once lost it cannot be found again by such Instruments. Nothing but Astronomy is sufficient for this purpose. But if you are unwilling to meddle with Astronomy - the only right method and the method pointed at by the Act of Parliament - I am unwilling to meddle with any other method than the right one.' [2]

Either the genius was right or wrong; there was no middle ground. Sobel also failed to realise that, again as Newton stated, the longitude prize had been set up (by him) so as to be won by astronomical means and a 'Clock-work' solution was never going to carry off the full prize. Nevertheless she wrote that Newton *'did not live to see the great longitude prize awarded at last, four decades later, to the self-educated maker of an over-sized pocket watch.'* [3]

Sometime in 1730, whilst the self-taught glazier Thomas Godfrey was still running sea trials with his marine angle-measuring device in the New World, (and John Hadley was yet to announce his 'invention' of a similar instrument), another artisan lacking any formal scientific training was making his move on the same longitude prize. A maker of wooden church clocks embarked on a 150-mile hike down the Great North Road to London from his village on Humberside to seek development finance for his sea-going clock project from the Board of Longitude commissioners.

Although Newton had repeatedly pointed out that longitude could not be obtained by 'clock-work' alone, Harrison was probably not aware of this and certainly would have known nothing of Newton's letters to the long-suffering Burchett. There is little doubt that Harrison originally genuinely believed that he would be entitled to the full elusive prize if he could but manufacture a reliable sea-going timer. He was presumably aware of the basic prize conditions; (Clause III appendix 7) half of the prize money to be paid when the commissioners were satisfied that the method (whatever it was) worked within 80 miles of a coastline and the other half after a successful oceanic sea trial from England to a West Indies port lasting at least eight weeks.

On arrival in the capital, Harrison was disappointed to discover that no member of the BOL was prepared even to see him. He travelled on to Greenwich where he confronted Edmond Halley; the Astronomer Royal was his last hope. Halley had little choice but to listen, but was soon swayed by the strength of Harrison's convictions. It is very difficult to believe that Halley of all people would not have drawn Harrison's attention to the small print in the Queen Anne prize rules; although he was at this particular time (1730) not in a position to discuss the important role an angle-measuring device would play in the longitude quest. Without being able to determine local time - by determining the exact time the sun reached its highest point in the sky with an angle-measuring device capable of being used from the deck of a ship on the high seas, a timer, however perfect, could not determine longitude; so any prize would have to be shared (with Newton's heirs possibly?).

Doubtless Halley had his reservations as to whether Harrison could make a sea-going timepiece anyway, but had the foresight to write him a letter of introduction to the one man he knew would help him; his Royal Society colleague, the instrument maker and horologist George Graham. Graham, to his everlasting credit, took Harrison under his wing, personally provided him with an unsecured loan for his project and stoutly defended him during his early efforts. Without the influential help of Halley and Graham, John Harrison's quest would surely have been doomed from the outset. Seven years later Harrison demonstrated his first marine timepiece; a somewhat larger affair than Halley had anticipated. Harrison's first attempt (H1) weighed in at a generous 34 kilograms!

By the time of Halley's death the BOL had advanced a total of £750 for research and development to Harrison, not specifically in order to help him win a portion of the Queen Anne prize but to assist in the development of a useful piece of navigational equipment.

Halley could so very easily have given Harrison the cold shoulder; even the man's surname must have evoked unpleasant memories and having been an interested bystander in the Hooke/Huygens chronometer fracas, surely enough was enough. As for Graham, a watchmaker with designs on a portion of the massive longitude prize himself; how many scientists would have helped and encouraged a potential rival as he did?

After Newton's death, a number of additions were made to the original Queen Anne Act. In 1741 the scope of the BOL research and development fund was widened (Geo II CAP 39 1741) to allow applications from surveyors seeking to determine the longitude and *latitude* of coastlines. By 1753, having now given Harrison a total of £1,250 from this fund, the BOL was left with only £250 in the kitty. As the commissioners had already promised Harrison more than this amount for work in hand, they had to return to parliament who provided a further £2000 for this fund (Geo II CAP 25 1753). However, Harrison's timepiece development was proving costly and by 1761 that £2,000 was also exhausted to the extent that the board was now in Harrison's debt to the tune of £300. In all Harrison was to receive a total of £1,815 from this R & D fund.

When Harrison's latest masterpiece (H4) was ready for testing, his son William left Portsmouth in November 1761 on board *Deptford*, a Royal Navy vessel en route for Jamaica. William and his father were by now firmly convinced that Halley's one-time assistant and successor, the 3rd Astronomer Royal James Bradley was placing every impediment in their path in order to scotch their efforts so as to win a portion of the prize himself. There may well have been some truth in this because the start of the voyage had been delayed more than once and setting off on a trans-Atlantic crossing in November would certainly be a harsh test for their precious chronometer.

When *Deptford* arrived in Port Royal, Jamaica, the BOL observer established local noon with a Hadley sextant and announced that H4 was only five seconds adrift from its agreed rate of change (Glossary). Well within the limits laid down in the rules. But when William arrived back in England the BOL commissioners refused to ratify the claim, pointing out that the longitude of Jamaica was not yet properly established so they had no proof that H4 had kept good time. Although this was hardly William's fault someone in the Harrison camp must surely have realised there was no point in testing a chronometer for its ability to confirm the longitude of Port Royal when no one was certain of the town's location. The same can be said for the commissioners of course and their observer should have accepted this responsibility; there would have been no problem over this as Jupiter was at a convenient altitude every night the pair were on the island.

However as H4 had completed a two-way Atlantic crossing and arrived back at its starting point only about 2 minutes adrift of its rate of change, the BOL could hardly refuse to admit the chronometer had performed satisfactorily, even if it had not exactly complied with the rules, and John Harrison was awarded £2,500 (£50,000) from the prize fund. Anticipating being awarded the full £20,000 prize, Harrison and his son, backed by the press and public who all suspected some sort of devious ploy on the part of the government, set about lobbying for a second trial.

At this point in time, May 1763, a leading French horologist Ferdinand Berthoud, accompanied by the astronomer Jerome de Lande and the mathematician Charles Camus paid a private visit to John Harrison. They had hoped to at least be shown H4, but the canny Harrison refused their request. If, as some reports suggested, a financial consideration was offered, most certainly this was also turned down. However, the Admiralty, well aware of the visit, now found themselves in a similar position to the one they had found themselves in following Newton's demonstration of his twin-mirrored octant to RS fellows in 1699. Without doubt, Harrison's chronometer was a national asset and unless something was done, the secrets of its construction could fall into enemy hands. Harrison would have to be placated.

Meanwhile predicting the lunar orbit was proving more difficult than anticipated. In 1719 Halley, using his hard-won knowledge had readied for the publisher, tables for calculating lunar positions linked to Newton's proposition as set out in the 1713 edition of *Principia*, but had delayed publication until he could check against observations he anticipated making at Greenwich. It was not until 1731, four years after Newton's death that he had announced he had completed those checks; his observations matched his predictions to better than 2' of arc, apparently confirming Newton's life-long quest.

Very little notice seems to have been taken of Halley's announcement, which implied that longitude could be established by the lunar distance method to within one degree by any navigator possessing a suitable angle-measuring device. Halley did not go out of his way to promote his claim because he was aware that not all was as straightforward as implied. The data should fit exactly, not 'almost', therefore either the data were wrong or the intricacies of the lunar orbit were not yet mastered (see below). He of all people knew that one must not encourage mariners to rely on 'almost' right predictions some of which had been obtained by hindsight.

After the death of Halley, the work of precisely logging and predicting the Moon's path round the Earth had been taken up by Bradley, who devoted 20 years to the task and came close to solving the puzzle. During his tenure a German astronomer, Johann Tobias Mayer produced a set of lunar tables giving the Moon's predicted location at regular intervals, based on observations taken at Gõttingen Observatory. In 1757, hoping for a share of the Queen Anne prize, Mayer sent his data to the First Lord of the Admiralty, Lord Anson, a BOL commissioner. Anson passed this data on to Bradley for evaluation. He checked these published predictions against his own actual observations made with the Bird mural quadrant, which had replaced the Graham instrument used by Halley. He pronounced the tables accurate to within ¾ of a degree of longitude (1½' of arc). Still not quite good enough, bearing in mind all the other little things that could go wrong with trying to take an accurate fix on the high seas. Whilst the commissioners were deciding on whether to make an award, Mayer died. His widow was eventually awarded the huge sum of £3,000 (£600,000) eight years later (see below); the same year Katherine Halley was awarded a paltry £100 for surrendering Halley's papers and journals.

Nevil Maskelyne was born in 1732 just as the dispute between Godfrey and Hadley over the marine octant invention was brewing. By all accounts, Mascelyne was yet another naturally brilliant mathematician and, like so many scholars of that era, he became an ordained minister of the Church. In 1758, he was admitted to the RS having been recommended because of his expertise in mathematics and natural philosophy. He was only 25 years old and already a fellow of Trinity College, Cambridge. He saw himself as a second Newton and in superficial ways, he was. The same Cambridge college, similar interests in mathematics and astronomy and elected to the RS at an even earlier age.

In 1762, Maskelyne published an updated version of the Mayer data, *'The British Mariners Guide'*, which in reality still contained most of the flaws noted by Bradley although this publication did firmly establish his reputation. When Bradley died soon after Mayer, his successor Nathaniel Bliss nominated Maskelyne to be the expert referee responsible for overseeing the next testing of Harrison's H4 chronometer. This time the voyage was to Bridgetown, Barbados and William Harrison again deputised for his father. Why H4 should have been granted a second sea trial when the first one had proved satisfactory can only be because either the commissioners hoped H4 would fail this time, or so as to encourage

Harrison not to sell his secrets to the French (or Spanish).

When H4 was checked for accuracy by Maskelyne and another astronomer Charles Green in Bridgetown in May 1764, they used a very expensive Bird-Hadley sextant fitted with a vernier device and then referred to a solar almanac in order to determine local noon. William Harrison then returned to England with his father's precious chronometer, leaving the longitude of Bridgetown to be determined by the referees using the Jovian moons method, a result that incidentally came very close to Halley's 1699 figure (appendix 4). Having then conferred with Bliss who had meanwhile established the longitude of the departure point by the same method, the observers eventually certified that H4 had, on arrival in Barbados, been in error by some 39 seconds against its pre-agreed rate of change; less than one sixth of a degree of longitude. The commissioners now of course found themselves in more of a quandary than ever. Exactly how were they to deal with a claimant whose instrument did not comply with the prize rules? A man whom they had already sponsored to the tune of £4,315 (£860,000) and who had devoted the best years of his life to the quest.

Before they could decide on a course of action, Nathanial Bliss died after only two short years at the helm. In January 1765, the Reverend Nevil Maskelyne was appointed the 5th Astronomer Royal, aged 33 although he had not been the only candidate for this plum appointment. Maskelyne may have jumped the queue by foregoing any rights to a portion of the Queen Anne prize he may have been anticipating, in return for Admiralty sponsorship of the post.

With Maskelyne now a member of the BOL, one of the first problems the commissioners faced in 1765 was how they were going to handle the inevitable public outcry when they suddenly admitted they were not empowered to award Harrison the longitude prize in full? Unable to pay because the device was far too expensive to manufacture (was not a *'practicable solution'* as stipulated in clause II of the act) and thus not of *'so great a use'* but only *'of considerable use'* (clause V). Worse, H4 could not accurately determine longitude without the help of a marine sextant. The commissioners, backs to the wall, were forced to return to parliament.

They publicly admitted their dilemma in the introduction to a supplementary act (Geo III, CAP 20; 1765) which also made provision to reward Mayer's widow for his lunar tables which were constructed *'upon the Principles of Gravitation laid down by* Sir Isaac Newton;...'. In referring to Harrison's claim the act stated; *'But Doubts may arise whether, by the Words of the said Act of the twelfth of Queen* Anne, *the Commissioners can direct the Payment of the said Reward of twenty thousand Pounds to the said John Harrison.'* The wording of this 1765 act then solved that problem officially by permitting the Navy to pay the full £20,000 to John Harrison in 2 parts, neatly removing Newton's invention from the competition; not that anyone was considering it. £10,000 would be awarded on handing over the first three timing devices H1, H2, & H3, revealing the inner workings of the H4 chronometer and surrendering it. The other £10,000 would follow *'So soon as other Time Keepers of the same Kind shall be made'* and successfully tested. Paragraph II of this act added another proviso; all the secrets had to be revealed within six months, otherwise Harrison was not entitled *any of the reward.* Forcing Harrison to hand over all four timepieces very neatly also solved the national security issue.

The sequence of the wording was unfairly and strictly adhered to which meant Harrison was forced to surrender H4 *before* being allowed to make *'other Time Keepers of the same Kind'* (note the plural). An impossible task for an old man and many were again incensed, including George III. Harrison hastily revealed *nearly* all and was awarded half the prize minus the £2,500 already awarded following the Jamaica fiasco. H4 had to date cost the Royal Navy £14,315 (£2.6 million).

Maskelyne then published his first volume of *'Nautical Almanac and Astronomical Ephemeris'* based on Mayer's tables and containing many pages of instructions. Had he not foregone his rights, Maskelyne would undoubtedly have been entitled to a portion of the £5,000 set aside under clause V of the new supplementary act for this specific purpose.

So by 1766 and within two years of Harrison proving his very expensive 'one-off' chronometer could keep good time at sea, the very cheap lunar almanac was in print and longitude on the high seas could at long last be determined by any highly competent navigator who could afford a twin-mirrored marine octant/sextant. Thus was the longitude at sea puzzle solved despite the long years of scientific infighting. The peculiarities of the Moon's orbit had finally been mastered, the advance prediction tables published and Newton's theory of universal gravitation confirmed; 39 years after his death. So much for Newton's claim to Flamsteed back in 1694 that if he could supply six or seven years of lunar data he could solve the puzzle within months. Within months? It had taken another seven decades to properly confirm because the Moon's motion was in fact not a straightforward 18+ year cycle after all, ~~but five~~ different inter-related cycles. For one, the miniscule gravitational push-pull effects of Jupiter, the original *sidereal messenger* had not been properly computed, which must have put a smile on the face of the king of the Olympian gods.

Newton's undertaking to his Maker had at last been fulfilled, but his wonderful marine octant invention, which should have received a portion of the main prize, had been sidelined during the interim. But why in that case did not John Hadley or his heirs receive an award when Jessie Ramsden had received £615 (chapter 30) merely for an improvement. The obvious answer is that the Admiralty who controlled the BOL purse strings knew full well Hadley was not the inventor (chapter 29).

An ageing and disillusioned Harrison assisted by his son, then produced a duplicate of H4 (H5), which took over four years to complete, and he clearly was never going to be able to make another. The king had H5 tested at his private observatory in Richmond Park, pronounced it perfect and pressured parliament into making a special award to Harrison of the 'missing' £10,000 minus the first three early R & D grants. The octogenarian was awarded £8,750 in June 1773. The final cost to the Royal Navy was £23,065 in return for five wonderfully innovative timepieces which if ever put on the market would fetch at least £100 million today. H4 and H5 were locked away, presumably for fear of the secrets of their inner workings falling into enemy hands. Again shades of Newton's octant and Halley's data. All five of John Harrison's marine timekeepers survive; four are marvelled at by thousands daily at the Royal Observatory at Greenwich and H5 is in the possession of the Worshipful Company of Clockmakers.

Although the story of John Harrison's attempt on the Queen Anne longitude prize is not exactly straightforward, it is clear the only prize he actually won outright was the King George III special £10,000 award; a competition only he had been allowed to enter.

Whilst Harrison was working on H5, the Admiralty had arranged for a replica of H4 to be manufactured by Larcum Kendal. It was this chronometer designated K1 which was tested by James Cook on his second voyage of discovery, which commenced in 1772 and which lost over 19 minutes against its rating in just less than a year. K1's performance was far inferior to that of H4 - owing to Kendal not being fully acquainted with Harrison's secrets - and Cook had to correct it (or rather he amended its rate of change) by determining his longitude by the lunar distance method just as Newton had predicted. Following Cook's exploits, it was not Halley but Dampier, who had never once been able to properly determine his longitude either at sea or on land, who was increasingly referred to as 'the Captain Cook of an earlier age.'

However, *technically* the longitude at sea question remained unresolved; heavy overcast prevented either latitude or longitude from being determined. The problem was eased considerably in 1913 when the United

States of America began broadcasting time signals more or less world-wide and consigned the tedious lunar distance method to the rubbish bin. No need to check the accuracy of the watch any more except in unusual conditions.

One such well-documented set of unusual circumstances serves as a bleak reminder of the frailty of marine chronometers. Frank Worsley the captain of *Endurance* and navigator to the ill-fated Shackleton 1914-1916 Antarctic expedition had no less than 24 marine chronometers with him when the ship left London in August 1914. By the time they arrived at Elephant Island in three open boats in April 1916, having had to abandon the sinking ice-trapped *Endurance*, Worsley was left with only one that was in good going order. As Worsley wrote shortly before he, Shackleton and four others set out for South Georgia on their epic voyage to seek help for the 22 men left behind, *'Immediately after breakfast the sun came out obligingly. The first sunny day with a clear enough horizon to get a sight for rating my chronometer'.*[4] Even though he already knew the exact position of Elephant Island and of South Georgia, Worsley had to check the rate of change of his watch before he dare navigate through the 600 miles of the dangerous Antarctic waters. Without his trusty twin-mirrored sextant (possibly the actual instrument now exhibited at the National Maritime Museum at Greenwich) to check his last reliable chronometer by astro-observation, all 28 expedition members could so easily have perished.

Yet the determination of longitude at sea in any weather was still not possible until the USA launched their series of orbiting satellites linked to ground-based receivers (GPS) in 1973. And just who was partly responsible for that? Halley and Newton (a very small contribution from Halley and a rather larger one from Newton), without whose mathematics and gravitational discoveries, the satellite launch trajectories could not have been computed. Newton's bald statement; *'Clockwork may be subservient to Astronomy but without Astronomy the longitude is not to be found'* had stood the test of time. Those satellites may not be as distant as the Moon or stars but nevertheless they are astronomical objects.

(1) Margaret Halley's age is noted incorrectly. She was baptised on 11th May 1685.

(2) R. Hall & L.Tilling, eds., *The Correspondence of Isaac Newton* (Cambridge, 1976), Vol. 6, p. 212.
(3) D. Sobel, *Longitude; the true story of a lone genius who solved the greatest scientific problem of his time* (London 1996), p.60.
(4) Frank Worsley, *Shackleton's Boat Journey* (London, 1999), p.101.

APPENDICES

APPENDIX 1

TEXT OF FLAMSTEED'S *'STATE OF THE OBSERVATORY'* NOTES, October, 1700.

Several persons, about the year 1674, pretending to the discovery of the longitude, and the most skilful of them proposing to find it by comparing the moon's apparent places (got by observing her distances from fixed stars) with her places given by astronomical tables (A), it was represented to his then Majesty, King Charles II, (by the Lord Brouncker, at that time principal officer at the Navy Board, Sir Jonas Moor, Surveyor-General of the Ordnance, and several other able mathematicians about the court) - that this method was indeed the most likely to prove useful to our sailors, because most practicable; but that the catalogue of the fixed stars made by Tycho Brahe, a noble Dane, an age agone, and now used, was both erroneous and incomplete:- that the best tables of the moon's motions (which, with the places of the fixed stars, must necessarily be employed in the enquiry of the longitude by this method) erred sometimes above 20 minutes; which would sometimes cause a fault of 15 degrees, or 300 leagues in the determination of the longitude by it:- that the longitudes of the coasts in our sea charts having been laid down from coarse accounts of sea voyages of our first navigators, and not from celestial observations, as they ought, were very erroneous; so that our sailors could expect no help from this method, till both the places of the fixed stars were rectified, and new tables of the moon's motions made, that might represent her places in the heavens to some tolerable degree of exactness; for which, a large stock of very accurate observations, continued for some years, was altogether requisite, but wanting: and that therefore his Majesty would give a great and altogether necessary encouragement to our navigation and commerce (the strength and wealth of our nation) if he would cause an Observatory to be built, furnished with proper instruments, and persons skilful in mathematics, especially astronomy, to be employed in it, to take new observations in the heavens, both of the fixed stars and planets, in order to correct their places and motions, the moon's especially; that so no help might be wanting to our sailors for correcting their sea charts, or finding the places of their ships at sea.

Hereupon his Majesty was pleased to order an Observatory to be built in Greenwich Park (B): Mr Flamsteed was appointed to the work, with the allowance of only £100 per annum, payable out of the office of the Ordnance; and a labourer in ordinary from the Tower, to move the instruments, count the clock, and call him at hours in the night proper for his business (C).

Notes on the above, by Mr. Flamsteed.

(A). This gentleman was a Frenchman, called himself Le Sieur de St. Pierre, and by an interest in the Duchess of Portsmouth, got himself recommended to the King

Charles II, who gave a commission to the Lord Viscount Brouncker, the Bishop of Salisbury (Dr. Ward), Sir Robert Morray, Sir Charles Scarborough, Sir Jonas Moor, Dr. Pell, and other eminent mathematicians about the Court, to hear his proposals. They met at Col. Titus's house; where, by a power given in their commission, I was chosen into their number, furnished him with such *data* as he required, and showed the insufficiency of them for the end he proposed; and suggested to them what they represented to the King, which they apprehended very easily.

(B). Chelsea College was proposed by some persons; and I went to see it. But, Greenwich Hill being mentioned by Sir Christopher Wren, the King approved of it, as the most proper; the same having been proposed to King Charles I, as I was told by Mr. Moore, an ingenious old mathematician then living: and that the Observatory was to have been built on the other hill in the park p.187, (chapter 6).; and an instrument, as large as any the Arabs boast of, fixed in it for determining the meridional heights and declinations of the fixed stars, and other astronomical observations.

(C). The laborer being paid by the Office of the Ordnance as well as myself, looked upon himself as a King's servant: and being a person only fit for hard labor, was rather a hindrance than help to me; but always a certain charge. For, I was forced to allow him diet; or want his attendance when I had occasion for him, on his pretence of providing it. Till, in the year 1694, the officers allowed me to name my own laborer: since which time I have named one of my own servants, and received his pay for his maintenance: which is a favor I must ever acknowledge.

Reference. Francis Baily, *An Account of the Revd. John Flamsteed* (London 1835/1966),

APPENDIX 2

ASTRONOMICAL OBSERVATIONS MADE BY EDMOND HALLEY, 1675-84

*Link to astronomical events * mentioned in chapters 6 & 7.*

Lunar occultations visible from Greenwich for the year 1675 should anyone have been observing.

Details of the 20 Lunar occultations including one case of three occultations in same night and five close appulses ¶ (including Mars). Stars all referred to by Flamsteed numbers.

11th January 19h.	85 Geminorum, magnitude 5.38 (during lunar eclipse)
12th January 21h.	62 Cancri, mag. 5.22
14th January 02h 30m.	29 Leonis, mag. 4.68. (just touching Moon)¶
4th February 00h 1m.	23 Tauri, mag. 4.16 (Hyades cluster)
4th February 00h 30m.	25 Tauri, mag. 2.85 (Hyades)
4th February 01h.	27 Tauri, mag. 3.61 (Hyades)

9th February 23h 30m. 5 Leonis, mag. 4.97
10th February 05h. 14 Leonis, mag. 3.52
9th March 23h 30m. 29 Leonis.
1st April 22h 30m. Mars (appulse)¶
6th April 01h 30m. 14 Leonis.
8th April 02h 40m. 87 Leonis, mag. 4.76
3rd June 23h 30m. 69 Virginis, mag. 4.76
6th June 23h. 21 Scorpii, mag. 0.88 (appulse)¶
13th June 04h. 43 Aquarii, mag. 4.17 (appulse)¶
26th June 21h. 14 Leonis (appulse - new Moon)¶
13th Aug 00h 00m. 47 Arietis, mag. 5.8
2nd September 02h. 18 Aquarii, mag. 5.48
17th September 05h 30m. 29 Leonis.
2nd October 01h. 8 Piscium, mag. 4.87
3rd November 06h 30m. 58 Arietis, mag. 4.87
5th December 04h 30m. 56 Geminorum, mag. 5.10
10th December 02h. 87 Leonis.
21st December 19h 30m. 63 Aquarii, mag. 5.04
31st December 00h 30m. 1 Geminorum, mag. 4.18.

11th January 1675.
Total lunar eclipse observed in London. Duration of totality; 18h 32m to 20h 09m. During the eclipse the 6th magnitude star 85 Geminorum was occulted by the Moon (see above).

23rd June 1675.
Annular partial solar eclipse at Greenwich; 51°28'40"N. Eclipse, which was not observed owing to cloud, commenced when Sun was still below the horizon at 03h 36m and ended at 05h 21m with the Sun at an altitude of 12°. Maximum eclipse 67%.

11th June 1676.
Annular partial solar eclipse observed at Greenwich. Eclipse commenced at 07h 50m with the Sun at an altitude of 35°. Maximum eclipse 33% at 08h 50m which ended at 09h 56m at an altitude of 53°.

17th May 1677.
Partial lunar eclipse observed on the island of St. Helena; 15°58'S., 05°45'W. Maximum eclipse observed at 03h 09m.

24th September 1678.
45 (rho^1) Sagittari eclipsed by Moon at 19h 06m, emerging at 20h 14m. 44 (rho^2) Sagittari was very close to the Moon at 20h, two hours after sunset.

29th October 1678.
Total lunar eclipse observed at Greenwich. Totality commenced at 19h 15m and ended at 20h 56m.

5th June 1679.

Lunar eclipse of Jupiter observed by Halley at Danzig. 54°22'N., 18°41'E. At 04h 15m (local time) eclipse of Jupiter by a crescent Moon commenced at 19° altitude. Jupiter emerged at 05h 15m with the Moon at 27° altitude. The Sun had recently risen (03h 30m) in the same part of the sky during the event. At Greenwich the eclipse commenced at 02h 56m when the Moon was at 7°37' altitude. Jupiter emerged at 03h 47m with the Moon at an altitude of 15°20'. Sunrise was at 03h 45m, making the event easier to observe than at Danzig.

24th June 1679.

A double lunar occultation was observed by Halley at Danzig when 45(rho^1) and 44(rho^2) Sagittari were eclipsed almost simultaneously at 23h 07m local time. Observation from Greenwich of this event would have provided useful information on the effects of parallax. Danzig's longitude was 75 minutes in time ahead of Greenwich but there was actually a 105 minute difference at the point which rho^1 was eclipsed at Greenwich. This difference was caused partly because the two observation platforms were separated by a little over 5% of the Earth's circumference and were viewing one close object and a pair of very distant ones, and partly because Greenwich is sited some 3° south of Danzig. Because of this difference of latitude, rho^2 was seen to be occulted from Danzig but at Greenwich merely brushed one limb of the Moon.

26th June 1679.

At 01h 00m local time in Danzig Halley observed three bright stars in Capricorn (5, 6 & 9 Capricorni) in line close above the Moon. This event could also have been observed from Greenwich.

1682-1684.

Early in the morning of 28th October 1682 the bright star Porrima (29 Virginis) was occulted by a crescent Moon. In all there were at least 50 other good opportunities to observe occultations and similar events by Halley from his private observatory in Islington, London during this period. There were no occultations of 15 Virginis (Zaniah) but on several occasions this star came sufficiently close to the Moon to enable Halley to obtain useful parallax and angular data which would serve him in good stead on 14th December 1699 (see chapter 19).

APPENDIX 3

TEXT OF ADMIRALTY SAILING INSTRUCTIONS TO HALLEY, 1698-1699.

First and second voyages of H.M.S.Paramore.

First voyage Admiralty's sailing instructions to Captain Edmond Halley, 15th (25th) October 1698 mentioned in chapter 16

Whereas his Maty. has been pleased to lend his Pink the Paramour for your proceeding with her on an Expedition, to improve the knowledge of the Longitude and variations of the Compasse, which Shipp is now compleatly Man'd, Stored and Victualled at his Mats. Charge for the said Expedition; you are therefore hereby required and directed, forthwith to proceed with her according to the following Instructions.

You are to make the best of your way to the Southward of the Equator, and there to observe on the East Coast of South America, and the West Coast of Affrica, the variations of the Compasse, with all the accuracy you can, as also the true Scituation both in Longitude and Latitude of the Ports where you arrive.

You are likewise to make the like observations at as many of the Islands in the seas between the aforesaid Coasts as you can (without too much deviation) bring into your course: and if the Season of the Yeare permit, you are to stand soe farr into the South, till you discover the Coast of the Terra Incognita, supposed to lye between Magelan's Streights and the Cape of Good Hope, which coast you are carefully to lay downe in its true position.

In your returne home you are to visit the English West India Plantations, or as many of them as conveniently you may, and in them to make such observations as may contribute to lay them downe truely in their Geographicall Scituation And in all the Course of your Voyage, you must be carefull to omit no opportunity of Noteing the variation of the Compasse, of which you are to keep a Register in your Journall.

You are for the better lengthning out your Provisions to put the Men under you Command when you come out of the Channel, to Six to four Mens Allowance, assureing them that they shall be punctually pay'd for the same at the End of the Voyage.

You are dureing the Term of this Voyage, to be very carefull in conforming your selfe to what is directed by the Generall Printed Instructions annex'd to your Commission, with regard as well to his Mats. honor, as to the Government of the Shipp under your Command, and when you returne to England, you are to call in at Plymouth and finding no Orders there to the contrary, to make the best of your way to the Downes, and remaine there till further Order. Giving Us an Accot. of your arrivall. Dated etc 15 Octor. 98.

HP. JH: JK. GW By &c. J.B.

To Captn. Edmd. Halley Commandn of his Mats. Pink the Paramour-River.
(Minuted:) Instructions for proceeding to Improve the knowledge of the Longitude and Variations of the Compasse.

Author's note.

J.B; Josaiah Burchett, Secretary to the Admiralty. GW; Goodwin Wharton.
HP; Henry Priestmen. JH; Sir John Houblon. JK; James Kendall.

Second voyage Admiralty's sailing instructions to Captain Edmond Halley, 12th (22nd) September 1699 mentioned in chapter 18.

Whereas his Majesty has been pleased to lend his pink ye Paramore, for your proceeding a second time wth her on an Expedition to Improve ye knowledge of the Longitude and variation of ye Compass, which ship is now Compleatly mann'd, stored and victualled at his Majestys Charge for ye said Expedition you are therefore hereby required and directed forthwith to proceed with her according to ye following Instructions.
You are without loss of time to sett saile with her and proceed to make a Discovery of ye unknowne southlands between ye Magellan Streights and ye Cape of good hope between ye Lattd of 50 and 55 South, if you meet not with ye land sooner Observing ye variation of ye Compass with all ye accuracy you can as also ye True Scituation both in Longitude & Lattd, of ye ports where you arrive.
You are likewise to make ye like observations at as many of ye Islands, in ye seas between ye aforesaid Coasts as you can (without too much deviation) bring into your Course.
In your returne home you are to visit ye English West India Plantations, or as many of ym as conveniently as you may, & in them to make such observations as may contribute to lay them downe Truly in their Geographical Scituation in ye Course of your Voyage, you must be careful to omitt no opportunity of noting ye variation of ye Compass, of which you are to keep a Register in your journal.
Your are for ye better lengthening out your provisions, to put ye Men under your Comand, when ye come out of ye Channell, to six to four Mens allowance assuring ym that they shall be punchially paid for ye same at ye End of ye Voyage.
You are dureing ye Terme of this Voyage, to be very carefull in conforming your self to what is directed by ye Generall printed Instructions annexed to your Commission, with regard as well to his Majesty's honour as to ye Government of ye ship under your Comand; and when you returne to England, you are to call in at Plymouth, & finding no orders there to ye contrary, to make ye best of your way to ye Downes & remaine there till further Order. Dated &c; 12 Sber 1699
JE: H: DM: By &c JB

To Captain Edmund halley Comand: of his Matys. Pink ye Paramour at Deptford.
(Minuted:) Instructions for his proceeding a second time to improve ye knowledge of the Longitude and Variation of ye Compass.

Author's note.
JE; John Egerton. H; Lord Haversham. DM; Sir David Mitchell. JB; Josaiah Burchett.

APPENDIX 4

ASTRONOMICAL OBSERVATIONS BY HALLEY, 1699-1701.

*Astronomical events ** and positional fixes mentioned in chapters 17,18,19 & 21*

A. Astronomical events mentioned in chapter 17.

<u>25th February 1699 – Moon appulse involving a star on the high seas.</u>
Location 03°26'S., and approximately 30°30'W.
Halley provided the following information:-
'This morning I observed the moon apply to a starr in fascia boreali [sagittari sign] *and concluded myself 160 leagues* [480 miles] *more westerly than our account, and but 50* [150 miles] *leagues to the East of Fernando Loronha.'*
The star Halley was referring to was Rho[1], a 3.90 magnitude star in Sagittarius which was in line with a crescent Moon before dawn (approx. 04h 30m local time) on 25th February 1699. This star was indeed in the middle of 'fascid boreali', the Milky Way and in the Sagittarius constellation. Halley eventually increased his DR longitude position by eight full degrees to 29°24'W., once he had found the time to work on the complex mathematics. His actual position was approximately 30°15'W. at the time.

<u>7th March 1699 - Moon appulse involving the star Aldebaran.</u>
Actual position 07°06'S., 35°00'W.
'On the night we fell in with the Coast viz Februr 25th [7th March] *I observ'd the Moon to apply to the Bulls Eye and that the starr was in a right line with the Moons horns when it was 10 deg 26 min high in the West, or at 10h 11' 44"* [pm estimated local time] *from both which observations I conclude that the Longitude of this Coast is a full 36 deg to the Westward of London wherefore we have been set by the currents to the Westwards, during the long calms, not less than 200 leagues.'*
The Moon and very bright star Aldebaran in the Hyades cluster (but known to Halley as Parilicium or the 'Bull's Eye') would have been in alignment later than the 22h 11m 44s as estimated by Halley. Partly due to a possible error in retaining local noon time, Halley's longitudinal assessment was almost exactly one degree in error. The first occasion seconds of time was mentioned by Halley.

<u>15th March 1699 - A lunar eclipse.</u>
Actual position 07°06'S., 34°53'W.
'...the End of the Eclipse of the Moon March 5th [15th] *in the evening - which as well as I could discern it, through the Clouds ended at 6h 17m; and the Moon appearing clear at 6h 22 1/2m was in penumbra so thick that I could not judge the Eclipse ended above 5 or 6 minutes.'*

<u>24th April 1699 - Jupiter eclipse of Io.</u>
Position 13°06'N., 59°37'W. (?) Bridgetown, Barbados.

'...found that at 11h.50' after noon, the Satellite [Io] *was very small as beginning to disappear..'*
Halley noted that if this event had been observed in Europe the longitude of Barbados could be properly determined. *'..but supposing - Mr Cassinis Tables true .... the longitude of the Island of Barbadoes from London will be 59°.5'.*
Halley does not mention the actual location of his observation and all one can say with certainty is his longitude was out by a maximum of 32'.

2nd June 1699 - Moon appulse involving 14 (omicron) Leonis (mag. 3.51) on high seas.
Approximate position 34°33'N., 62°30'W.
'This Evening I observed the Moon to apply to a Starr in the foot of Leo.'
There was no alteration in the longitude details written in Halley's personal log and whilst the star can be identified as 14 Leonis, the lack of any other information precludes any precise assessment of his conclusions.

22nd June 1699 - Moon appulse involving 98 (mu) Piscium (mag. 4.84) on high seas.
Approximate position 48°00'N., 23°00'W.
'in the Morning the Moon apply'd to a Starr in Line piscium by which I find my Self 25 leagues more westerly than my Reckoning.'
Halley noted this 'in line' (appulse) fix put him 1½° short of his DR position but again, lack of information precludes any precise assessment of his conclusions. However an examination of his unaltered log, which places the Scilly Isles 3° out of position based entirely on DR would suggest that his longitudinal mid-ocean fix was about 1° in error and he was at that point about 2½° short of his DR position.

B. Astronomical events referred to chapter 18.
During the interval between voyages (25th July to 25th September 1699), Halley observed occultations in London. He could have observed six, five of which would have been in September.
19th August. 87 (Aldebaran/Parilicium) Tauri (Hyades).
11th September. 98 Piscium.
15th September. 54 Tauri (Hyades).
15th September. 75 & 77 Tauri at same time (Hyades).
16th September. 111 Tauri (outside the Hyades cluster).
18th September. 54 Geminorum.

C. Astronomical events referred to in chapter 19.
2nd November 1699 - Jupiter -Ganymede transit observation.
Actual position 16°42'N., 22°55'W.
At 18h 32m local time the largest of Jupiter's moons Ganymede suddenly became visible as it emerged from transit. From this single observation Halley was able fix the longitude of the island of Sal in the Cape Verde Group as 23°W. He also determined the island's latitude correctly as being between (NE end) 16°55'N. and (S end) 16°35'N.

<u>14th December 1699 - Moon appulse involving star 15 Virginis (eta/Zaniah) on high seas.</u>
Approximate position 20°45'S., 34°25'W.
'This Morning the Moon Aplyed to a Starr in Virgo of the 4th Mag. whose Longitude is [Libra sign] *0deg 39' Lat 1 deg 25'. The Moon did exactly Touch this Starr with her Southern Limb at 3h.15' in the Morning and at 3h 20' 20" the Southern horn was just 2 Minutes past the Starr haveing carefully examin'd this observation and Compared with former observations made in England I conclude I am in a True Longitude from London at the time of this observation 36deg15' and at this noon 36deg 35'. That is according to the Accot I have of it, about 5 Degrees East of Cape Frio.'*
This paragraph is probably one of the most important 'longitude at sea' statements ever made. The star in question was Zaniah. There are minor errors in the star's cited position but these are of no consequence. This is the second and only other occasion Halley mentions time to the nearest second in any of his journals. Importantly he mentions bringing data with him (for the only time), although obviously he always did. Data on Zaniah would have been collected from London (chapter 7 & appendix 2) but not during Halley's 1699 vacation (chapter 18) as the star was not anywhere near the Moon during that period. Position probably a little over 1° short of Halley's estimate.

<u>15th December 1699 - Moon and Mars in line - on high seas.</u>
Approximate position. 21°30'S., 35°15'+W.
'The Moon apply'd to Mars who was in line wth her horns at 4:3 or when CorM was 8°:6' high in the East.'
'CorM' was almost certainly a mistake of the copier and referred not to Corvus m, but to 'Scor M', mu Scorpii which were the twin stars mu^1 & mu^2 and very bright in 1699. The time is also confusingly abbreviated; 4:3 when meaning 04h 30m. These two stars were actually at an altitude of about 8°12', so Halley's very accurate angular measurement between horizon and stars, taken at night and having to make some allowance for refraction was only possible with Newton's twin-mirrored instrument. Subsequent dead reckoning navigation to a named landfall five days later suggests that the two longitudinal fixes of 14th and 15th December were both about 1° in error.

<u>21st April 1700 - Moon and Hyades cluster observation on high seas.</u>
Approximate position. 20°25'S., 29°00'W.
On 21st April, Halley managed to obtain his first lunar fix for 17 weeks. *'Last night the Moon Apply'd to the Contiguae in facie Tauri and I got a very good observation, whence I conclude my Selfe 2 degrees more to the Westward than by my Account* [of 27°20'W.].*'*
Halley was at most only 30' in error and was referring to the Hyades cluster in the Taurus constellation. The new Moon was in amongst them and Halley was familiar with the group's position having obtained four very good sets of positional plots in London whilst between trips - see chapter 18 notes above.

<u>9th August 1700 - Moon and Hyades cluster observation on high seas.</u>
Approximate position. 44°22'N., 56°W.
'This morning I observed the Moon aply to ye Hiades.'
From this second fix on the Hyades cluster Halley deduced his longitude near Cape Pine at the southern tip of St Mary's Bay, Newfoundland to be 53°53'W. (actually 53°10'W.) and adjusted his DR longitude slightly. The best fixes would have come as dawn was breaking.

Longitudinal assessments.
Final column represents the number of days since obtaining a land-based positional fix (at the point of departure on the first voyage; Greenwich and on the island of Sal; second voyage).

First Voyage.

Date	*Method*	*Location*	*Accuracy*	*Days since fix*
25th Feb 1699	Lunars	2 days from land	- 0°45'	66
7th March 1699	Lunars	near coast	+ 1°	78
24th April 1699	Jupiter	ashore	- 0°30'	
2nd June 1699	Lunars	in open ocean		
22nd June 1699	Lunars	in open ocean	+ 1°?	

Second Voyage.

2nd Nov 1699	Jupiter	ashore	+ 0°05'	
14th Dec 1699	Lunars	in open ocean	+ 1°?	
15th Dec 1699	Moon Mars	in open ocean	+ 1°?	
21st April 1700	Lunars	3 days from land	+ 0°30'	170
3/16 July 1700	Jupiter	ashore	- 1°	
9th Aug 1700	Lunars	2 days from land	+0°45'	

Latitudinal assessments.
In addition to the 11 longitudinal fixes, during Halley's three voyages, *latitudes* were established by taking sightings of the height of the noon Sun on a number of *stated* occasions (by *'observation'*) - Nine times on the first voyage, 156 on the second and 15 on the short third. Obviously many more were taken on the first voyage but not noted as such in Halley's journal. Only when these observations could be matched to a known landmark are they detailed below. Latitudes could only have been obtained with this degree of accuracy using a sophisticated angle-measuring device (figure 13 chapter 30) capable of measuring angles to 90°. The relatively poor result obtained at Funchal was probably due the extreme weather conditions.

First Voyage.

Location	*Halley latitude*	*Actual latitude*	*Naut. Mile error*	*Noon Sun altitude*
Funchal	32°30'N	32°38'N	8	34°
Sal	16°40'N	16°36'N	4	51°
Fernando Noronha	3°57'S	3°52'S	5	86°
Pariba river	7°00'S	6°58'S	2	84°
Barbados	13°10'N	13°05'N	5	87°
Scilly (North)	49°57'N	49°59'N	2	63°

Second Voyage.

Location	*Halley latitude*	*Actual latitude*	*Naut. Mile error*	*Noon Sun altitude*
Heirro	27°45'N	27°45'N	0	49°
Sal	16°35' -55'N	16°36' - 53'N	0	58°
Boa Vista	16°00'N	16°00'N	0	58°
Nightingale	37°25'S	37°28'S	3	60°
Jamestown St.Helena	15°52'S	15°55'S	3	73°
Trinidade	20°25'-29'S	20°25'-31'S	0	56°
St. George's	32°24'N	32°24'N	0	81°
Scilly (South)	49°50'N	49°50'N	0	46°

1700 Channel survey voyage (Chapter 21).

Beachy Head	50°45'N	50°44'N	1	61°
Bonchurch IOW	50°32'N	50°36'N	4	60°
Cap de la Hague	49°46'N	49°44'N	2	60°
Alderney	49°47'N	49°44'N	3	60°
Portland Roads	50°32'N	50°34'N	2	58°
Plymouth	50°20'N	50°21'N	1	53°
Eddystone Light	50°08'N	50°10'N	2	53°
Ushant	48°28'N	48°28'N	0	52°
Start Point	50°10'N	50°13'N	3	48°
Cape Frehell	48°41'N	48°41'N	0	45°
Les Sept Isles	48°57'N	48°53'N	4	44°
St Aubin Jersey	49°15'N	49°12'N	3	43°

APPENDIX 5

TEXT OF HALLEY'S LETTER TO BURCHETT FROM BERMUDA, 1700.

Letter sent by Halley to Josaiah Burchett, Admiralty Secretary from Bermuda 8th (19th) July 1700 referred to in chapter 19 .

Bermudas July 8° 1700
Honourd Sr
My last from St Hellena, gave your Honour an account of my Southern Cruise, wherin I endeavoured to see the bounds of this ocean on that side but in the Lattd. of 52½° was intercepted with Ice cold and foggs Scarce credible at that time of the Year. Haeving spent above a Month to the Southwards of 40 degrees, and Winter comeing on, I stood to the Norwards again and fell with the three Islands of Tristan da Cunha which yielding us noe hopes of refreshment, I went to St. Helena, where the continued rains made the water soe thick with a brackish mudd, that when settled it was scarce fitt to be drunke; all other necesarys that Island furnishes abundantley. At Trinidad we found excellent good water, but nothing else. Soe here I changed as much of my St. Hellena water as I could, and proceeded to Fernambouc in Brassile, being desirous to hear if all were at peace in Europe, haveing had noe sort of Advice for near eight months, here one Mr. Hardwyck that calls himselfe English consull, shewed himselfe very desirous to make prize of me, as a pyrate and kept me under a guard in his house, whilst he went aboard to

examine, notwithstanding I shewd him both my commisions and the smallness of my force for such a purpose, from hence in sixteen days I arrived at Barbados on the 21st of May, where I found the Island afflicted with a Severe Pestilentiall dissease, which scarce spares any one and had it been as mortall as common would in a great measure have Depeopled the Island. I staied theire but three days, yet my selfe and many of my men were seazed with it, and tho it used me greatly and I was soon up again yet it cost me my skin, my ships company by the extraordenary care of my Doctor all did well of it, and at present we are a very healthy ship: to morrow I goe from hence to coast alongst the North America and hope to waite on their Lordsps: my selfe within a month after the arrivall of this, being in great hopes, that the account I bring them of the variations and other matters may appear soe much for the public benefit as to give their Lordsps intire satisfaction:
I am Your Honrs most Obedt Servant:
Edmond Halley

Notes.
1.It has since been suggested that the 'Pestilentiall dissease' was either yellow fever, typhoid fever or an African gastro-intestinal disease imported with the slaves.
2. The letter and signature were not in Halley's autograph.
3. The two most important events, the claiming of Trinidade for the Crown and the numerous longitudinal plots obtained, were not specifically mentioned.

APPENDIX 6

TEXT OF HALLEY'S 'WARNING TO MARINERS' PUBLICATIONS 1701.
Dual text of an anonymous Royal Society publication (Phil. Trans. 1701, Vol. 22 pp.725-6) and subsequent anonymous broadside pamphlet, both in fact authored by Edmond Halley referred to in chapters 21 and 23.

The anonymous broadside pamphlet titled An ADVERTISEMENT Necessary to be Observed in the NAVIGATION Up and Down the CHANNEL of ENGLAND. Communicated by a Fellow of the Royal Society and published in 1701 also contains the full text of the RS publication which is virtually identical but ends 'fair by the *Lizard.*'

For several Years last past it has been Observed, that many ships bound up the Channel, have by mistake fallen to the Northward of *Scilly*, and run up the *Bristol Channel* or *Severn Sea*, not without great Danger, and the Loss of many of them. The Reason of it is, without dispute, from the Change of the Variation of the Compass, and the Latitude of the *Lizard* and *Scilly*, laid down too far Northerly by near 5 Leagues. For from undoubted Observation the *Lizard* lies in 49° 55', the middle of *Scilly* due west therefrom, and the South part thereof nearest 49° 50'.

whereas in most Charts and Books of Navigation they are laid down to the Northward of 50°: and in some full 50° 10'. Nor was this without a good Effect as long as the Variation continued Easterly, as it was when the Charts were made. But since it is become considerably Westerly, (as it has been ever since the year 1657) and is at present about 7 ½ Degrees; all Ships standing in, out of the Ocean, East by Compass, go two Thirds of a Point to the Northward of their true Course; and in every eighty Miles they Sail, alter their Latitude about ten Minutes; so that if they miss an Observation for two or three Days, and do not allow for this Variation, they fail not to fall to the Northward of their Expectation, especially if they reckon *Scilly* in above 50 Degrees. This has been by some attributed to the Indraught of St. *George's* Channel, the Tide of Flood being supposed to set more to the Northward, than is compensated by the Ebb setting out. But the Variation being allowed, it hath been found that the said Indraught is not sensible, and that Ships steering two Watches *E b S* for one *East*, do exactly keep their Parallel. This practice is therefore recommended to all Masters of ships, who are unacquainted with the Allowances to be made for the Variation; as also that they come in, out of the Sea, on a Parallel not more Northerly than 49° 40' which will bring them fair by the *Lizard.*

The RS publication ends at this point and the following material was gleaned by Halley from his Channel Survey (chapter 21).

Nor is this the only Danger to which Ships are exposed in the *Channel*, on account of this Change of the Variation; for this last Winter has given us more than one Instance of Shipwreck upon the *French* coast and the *Casketts* of Ships newly departed from the *Downs*. And though perhaps this were not the only Cause of these Losses, yet it cannot be doubted but it concurred thereto: For by the late curious Survey of the Coast of *France*; compared with what has been done for our own (though perhaps not altogether so exactly) it appears that the true Course from the Land of *Beachy* or *Dungyness* to the *Caskett*-Rocks, is but West 26 Degrees Southerly; which in former Times, when the Variation was as much Easterly, as now 'tis Westerly, was about *SW b W* by Compass, and then a *W S W* Course, then called *Channel Course*, was very proper for all Ships bound into the Ocean: But at present whoso steers a *W S W* Course in the *Channel*, though never so near to the Shore of *Beachy*, will not fail to fall in with the *Casketts*, or rather to the Eastwards thereof: It follows therefore, That as the Compass now Varies, the West by South Course be accounted the *Channel* Course, instead of *W S W*; which Course from a reasonable Offing from *Beachy-head* will carry a Ship fair without the *Isle of Wight*, and about mid-way between *Portland-Bill* and the *Caskett*-Rocks; which are scarce 14 Leagues asunder, and nearly in a *Meridian*. If this Notice be thought needless by those, whose Knowledge and Experience makes them want no Assistance; yet if it it may contribute to the saving of any one Ship, the Author thereof is more than recompenced for the little pains he has taken to communicate it.

LONDON. Printed for *Sam. Smith and Benj. Walford.* Printers to the Royal Society, at the *Princes Arms* in *St. Pauls* Church-Yard. 1701 Price Two Pence.

APPENDIX 7

TEXT OF LONGITUDE PRIZE RULES, 1713.
Queen Anne Act of Parliament Longitude prize conditions (chapters 25 & Epilogue).

A.D. 1713. CAP. XV

An Act for providing a Publick Reward for such Person or Persons as shall discover the Longitude at Sea.

Whereas it is well known by all that are acquainted with the Art of Navigation, That nothing is so much wanted and desired at Sea, as the Discovery of the Longitude, for the safety and Quickness of Voyages, the Preservation of Ships, and the Lives of Men: And whereas in the Judgement of able Mathematicians and Navigators, several Methods have already been discovered, true in Theory, though very difficult in Practice, some of which (there is Reason to expect) may be capable of Improvement, some already discovered may be proposed to the Publick, and others may be invented hereafter: And whereas such a Discovery would be of particular Advantage to the trade of *Great Britain*, and very much for the Honour of this Kingdom; but besides the great Difficulty of the Thing itself, partly for the Want of some Publick Reward to be settled as an Encouragement for so useful and beneficial a Work, and partly for want of Money for Trials and Experiments necessary thereunto, no such Inventions or Proposals, hitherto made, have been brought to Perfection; Be it therefore enacted by the Queen's most Excellent Majesty, by and with the Advice and Consent of the Lords Spiritual and Temporal, and Commons, in Parliament assembled, and by the Authority of the same, That the Lord High Admiral of *Great Britain,* or the First Commissioner of the Admiralty, the Speaker of the Honourable House of Commons, the first Commissioner of the Navy, the first Commissioner of Trade, the Admirals of the Red, White, and Blue Squadrons, the Master of Trinity-House, the President of the Royal Society, the Royal Astronomer of *Greenwich,* the *Savilian, Lucasian* and *Plumian* Professors of Mathematicks in *Oxford* and *Cambridge,* all for the Time being, the Right Honourable *Thomas* Earl of *Pembroke* and *Montgomery, Philip* Lord Bishop of *Hereford, George* Lord Bishop of *Bristol, Thomas* Lord *Trevor,* The Honourable Sir *Thomas Hanmer* Baronet, Speaker of the Honourable House of Commons, the Honourable *Francis Roberts* Esq; *James Stanhope* Esq; *William Clayton* Esq; and *William Lowndes* Esq; be constituted, and they are hereby constituted Commissioners for the Discovery of the Longitude at Sea, and for examining, trying, and judging of all Proposals, Experiments, and Improvements relating to the same: and that the said Commissioners, or any five or more of them, have full Power to hear and receive any Proposal or Proposals that shall be made to them for discovering the said Longitude; and in case the said Commissioners, or any five or more of them, shall be so far satisfied of the Probability of any such Discovery, as to think it proper to make Experiment thereof, they shall certify the same, under their Hands and Seals, to the Commissioners of the Navy for the Time being, together with the Persons Names, who are the Authors of such Proposals; and upon producing such Certificate, the said Commissioners are hereby authorized and required to make out a Bill or Bills for any such Sum or Sums of Money, not exceeding two thousand Pounds, as the said Commissioners for the Discovery of the said Longitude, or any five or more of them, shall think necessary for making the Experiments, payable by the Treasurer of the Navy, which Sum or Sums the Treasurer of the Navy is hereby required to pay immediately to such Person or Persons as shall be appointed by the Commissioners for the Discovery of the said Longitude, to make those Experiments, out of any Money that shall be in his Hands, unapplied for the Use of the Navy.

II. And be it further enacted by the Authority aforesaid, That after Experiments made of any Proposal or Proposals for the Discovery of the said Longitude, the Commissioners appointed by this Act, or the major Part of them, shall declare and determine how far the same is found practicable, and to what Degree of Exactness.

III. And for a due and sufficient Encouragement to any such Person or Persons as shall discover a proper Method for finding the said Longitude, Be it enacted by the Authority aforesaid, That the first Author or Authors, Discoverer or Discoverers of any such Method, his or their Executors, Administrators, or Assigns, shall be intitled to, and have such Reward as herein after is mentioned; that is to say, to a Reward, or Sum of ten thousand Pounds, if it determines the said Longitude to one Degree of a great Circle, or sixty Geographical Miles; to fifteen thousand Pounds, if it determines the same to two Thirds of that Distance; and to twenty thousand Pounds, if it determines the same to one Half of the same Distance; and that one Moiety or Half-Part of such Reward or Sum shall be due and paid when the said Commissioners, or the major Part of them, do agree that any such Method extends to the Security of Ships within eighty Geographical Miles of the Shores, which are Places of the greatest Danger, and the other Moiety or Half-Part, when a Ship by the Appointment of the said Commissioners, or the major Part of them, shall thereby actually sail over the Ocean, from *Great Britain* to any such Port in the *West Indies,* as those Commissioners, or the major Part of them, shall chose or nominate for the Experiment, without losing their Longitude beyond the Limits before mentioned.

IV. And be it further enacted by the Authority aforesaid, That as soon as such Method for the Discovery of the said Longitude shall have been tried and found practicable and useful at Sea, within any of the Degrees aforesaid, That the said Commissioners, or the major Part of them, shall certify the same accordingly, under their Hands and Seals, to the Commissioners of the Navy for the Time being, together with the Person or Persons Names, who are the Authors of such Proposal; and upon such Certificate the said Commissioners are hereby authorized and required to make out a Bill or Bills for the respective Sum or Sums of Money, to which the Author or Authors of such Proposal, their Executors, Administrators; or Assigns, shall be intitled by virtue of this Act; which Sum or Sums the Treasurer of the Navy is hereby required to pay to the said Author or Authors, their Executors, Administrators, or Assigns, out of any Money that shall be in his Hands unapplied to the Use of the Navy, according to the true Intent and Meaning of this Act.

V. And it is hereby further enacted by the Authority aforesaid, That if any such Proposal shall not, on Trial, be found of so great Use, as aforementioned, yet if the same, on Trial, in the Judgement of the said Commissioners, or the major Part of them, be found of considerable Use to the Publick, that then in such Case, the said Author or Authors, their Executors, Administrators or Assigns, shall have and receive such less Reward therefore, as the said Commissioners, or the major Part of them, shall think reasonable, to be paid by the Treasurer of the Navy, on such Certificate, as aforesaid.

APPENDIX 8

HYADES CLUSTER / LUNAR EVENTS, 1718-1720.

Relating to Edmond Halley's anonymous 1717 publication mentioned in chapter 18. Dates when the Hyades cluster (in the Taurus constellation) and the Moon could be used to determine longitude; January 1718 to December 1720 in, for example the West Indies (Barbados local times illustrated) and possibly secure a portion of the Queen Anne Longitude prize. That this advice was published anonymously and without any mention of Halley's own successes using this method, confirms he was still constrained by the Admiralty after the establishment of the Longitude prize.

<u>1718</u>
13th January. Moon's luminosity 82%. Local time 01h 00m. Moon's alt. 26°.
8th March. 38% - 19.30 - 56°.
4th April. 16% - 20.30 - 12°.
16th September. 65% - 02.30 - 58° - 87 Tauri (Aldebaran) occultation.
13th October. 86% - 02.30 - 86°.
6th December. 99% - 22.30 - 78° - 87 Tauri appulse.
<u>1719</u>
3rd January. 88% - 02.00 - 20°.
26th February. 46% - 22.00 - 27°.
6th September. 58% - 00.00 -14° - 75 & 77 Tauri occulted & 78 Tauri appulse.
3rd October. 80% - 05.00 - 65°.
26th November. 99% - 18.30 - 13° - 75 Tauri occultation.
24th December. 94% - 03.00 -17° - 63 Tauri occultation.
<u>1720</u>
16th February. 56% - 20.00 - 66°.
22nd September. 72% - 02.30 - 67° - 63 Tauri occultation.
15th November. 99% - 20.30 - 30°.
13th December. 97% - 00.00 - 70° - 64 Tauri occultation, 61 Tauri appulse.

APPENDIX 9

ARE THE CIPHERS GENUINE? *(chapters 24, 26, 28, 29 & Epilogue*)

In 1997 Michael Drosnin caused a furore with the publication of *'The Bible Code'* in which he detailed the claims of scientists who had discovered all manner of hidden predictions in The Bible. Their technique was to search for equidistant letter sequences (EDLS's); selecting a piece of text and a starting letter and then jumping forward a set number of letters at a time - a skip sequence. In this way they had 'discovered' Newton's name in a vertical line of Hebrew letters which was crossed by a horizontal line which included the word 'gravity'. As critics have pointed out if one selects a long enough slice of text, can vary the starting place and the size of the jump at will and can juggle with the horizontal, vertical and oblique angles, words and even meaningful phrases are sure to turn up especially when a computer is being used. In Drosnin's examples advantage had been taken of Hebrew texts largely devoid of vowels which enabled the insertion of these at suitable places. Significantly and unlike the Wren cipher, the computer generated EDLS's had no logical base. In reality there was no key and the claims were without statistical merit. Are the following any different?

The reason for first hidden message.

1707. The Shovell disaster.
The disaster was caused because data obtained by Newton's instrument was ignored by Shovell which could have triggered a response from the designer and/or user of the instrument.

The chronological sequence of the hidden messages.

1. 1708. The Shovell monument.
The key was based on numerology and was the sum of the numbers in the text.
The hidden message was signed.
Forward counting of capitalised words.
Comments.
The hidden message was brief and to the point and Newton had motive and opportunity to have influenced the layout and capitalisation of words. Halley could have been the culprit but using Newton's initials without approval was not his style. The four words ***'Deservedly Shipwreck'd Scilly In'*** represent the most insulting and appropriate sequence possible to construct. Two other sequences totalling 31 do not include the essential 'Scilly'.... *Sr Cloudesley Shovell Deservedly Shipwreck'd. In.*, and *....Admiral Deservedly Shipwreck'd. In.*

If the word 'Services' had not been provided with a capital letter, 'Scilly' could not have been incorporated in the hidden statement. It could be argued that the peculiar shape to the capital '*S*' in 'Services' indicates this was altered from lower case deliberately to this end.

<u>2. 1714. The Wren cipher.</u>
The smokescreen key was based on numerology.
The hidden key was based on numerology and was the sum of the numbers in the text.
Both hidden messages were signed.
Backward counting of all letters including Roman numerals.
Comments.
Wren wrote the letter and placed his name within the basic text. The hidden message was entirely relevant and Wren who had a direct interest wrote the text himself, sending it to a person he suspected already possessed the key.

<u>3. 1727 The Newton monument.</u>
The key was based on numerology and was the sum of the numbers in the text.
The hidden message was signed.
Backward counting of all words and abbreviated words excluding Roman numerals.
Comments.
The hidden message was entirely relevant and Halley who had a direct interest almost certainly wrote or assisted in the composition of the epitaph.

<u>4. 1731 The John Hadley cipher.</u>
The key was based on numerology and was the sum of the numbers in the text.
The message was signed.
Backward counting of capitalised words.
Comments. There were a large number of possible messages but only one was relevant and that was obvious.

<u>5. 1742 The Halley Memorial ciphers.</u>
The first was a repeat of the John Hadley cipher sum and method.
The key to the second was also based on numerology and was the sum of the numbers in the text.
Forward counting of all words.
Comments. The hidden messages were entirely relevant and Halley could have influenced the composition of his own epitaph.

<u>Collective supportive evidence.</u>
All the hidden messages are signed.
In each case the key is the sum of the individual numbers in the text and the resulting deciphered messages are all apposite.
Although the numbers key can be used in a variety of ways (forward or backward counting, to include or exclude the numerals or to count only words with capitals etc.), the reasons for these variations appear fairly obvious to anyone holding the respective key.

For example, Wren sent a hidden message to Newton (who held the key to the Shovell monument cipher) which was obviously meant to be read backwards and also hinted at using the numbers key three times. Halley had to write his hidden message backwards in order to make his initials fit. He then used a similar backward count in his Royal Society publication in order to make the statement read correctly (*No Invention* rather than *Invention No*) but used only capitalised words, mimicking the original Shovell example.

Taken individually each example could be nothing more than a series of strange coincidences. Therefore the underlying burden of proof seems to lie in whether or not those involved were each aware of the number-based key used previously. A key originated by Newton which reflected his long fascination and deep research into the subject of numerology which he had been using in his attempts to prove that The Bible contained a code which predicted dates for the second coming of Jesus and of Armageddon.

If Wren was not aware of Newton's presumed use of a specific type numerology-based key (the sum of each individual number) on the Shovell monument, there was no point in using the same specific method himself; Newton could *never* have deciphered it without the key.

But from Newton's subsequent behaviour it seems he succeeded. In which case Newton did use it on Shovell's monument (or someone conspired to do so on his behalf) and Wren had discovered this. Wren's hidden cipher was clearly intentional and not a mere coincidence, as evidence the otherwise pointless missing letter 'T' from the second string which prevented an 11 letter endless repeat cycle. Halley must also have known about the Shovell monument cipher because he subsequently used the same technique himself.

Should the reader ever have occasion to visit Westminster Abbey, glance up at the epitaph on the peculiar monument to Sir Cloudesley Shovell in a cul-de-sac opposite the Quire. Then, set in the wall, half way along the far (south) side of the Cloisters you will find a memorial plaque to Edmond Halley dedicated in 1986. Finally arriving at Sir Isaac Newton's monument in the nave, consider whether Halley really did include the hidden message *HUMANI GENERIS gratulentur E.H.*

When you visit the Royal Observatory at Greenwich, pause for a moment outside the Camera Obscura and read the inscription on the wall (Epilogue). Indoors marvel at Halley's mural quadrant but please don't touch the brass-work. Hopefully it may by now have been re-hung facing south.

APPENDIX 10

THE FULL TEXT OF NEWTON'S NOTE.
Discussed in Chapter 30, together with notes on the 5 serious errors contained in the published engraving (figure 12).

A. The text.
The full text [1] of Newton's note as published in the *Philosophical Transactions of the Royal Society, Vol.42, no. 465 pp 155-6, 1742* but with paragraph numbers added, reads as follows:-

1. In the annexed Scheme, PQRS denotes a Plate of Brass, accurately divided in the Limb DQ, into 1/2 Degrees, 1/2 Minutes, and 1/12 Minutes, by a Diagonal Scale; and the 1/2 Degrees, and 1/2 Minutes, and the 1/12 Minutes, counted for Degrees, Minutes, and 1/6 Minutes.

2. AB, is a Telescope, three or four Feet long, fixt on the edge of that Brass Plate.

3. G, is a Speculum, *fixt on the said Brass Plate perpendicularly, as near as may be to the Object-glass of the Telescope, so as to be inclined 45 Degrees to the Axis of the Telescope, and intercept half the Light which would otherwise come through the Telescope to the Eye.*

4. CD, is a moveable Index, turning about the Centre C, and, with its fiducial Edge, shewing the Degrees, Minutes, and 1/6 Minutes, on the Limb of the Brass Plate PQ; the Centre C, must be over-against the Middle of the Speculum *G.*

5. H, is another Speculum, *parallel to the former, when the fiducial Edge of the Index falls on 00d 00' 00"; so that the same Star may then appear through the Telescope, in one and the same Place, both by the direct Rays and by the reflex'd ones; but if the Index be turned, the Star shall appear in two Places, whose Distance is shewed, on the Brass Limb, by the Index.*

6. By this Instrument, the distance of the Moon from any Fixt Star is thus observed; View the Star through the Perspicil [telescope] *by the direct Light, and the Moon by the Reflext (or on the contrary); and turn the index till the Star touch the Limb of the Moon, and the Index shall shew upon the Brass Limb of the Instrument, the Distance of the Star from the Limb of the Moon; and though the Instrument shake, by the Motion of your Ship at Sea, yet the Moon and Star will move together, as if they did really touch one another in the Heavens; so that an Observation may be made as exactly at Sea as on Land.*

7. And by the same Instrument, may be observed, exactly, the Altitudes of the Moon and Stars, by bringing them to the Horizon; and thereby the Latitude, and Times of Observations, may be determined more exactly than by the Ways now in Use.

8. In the Time of the Observation, if the Instrument move angularly about the Axis of the Telescope, the Star will move in a Tangent of the Moon's Limb, or of the Horizon; but the Observation may notwithstanding be made exactly, by noting when the Line, described by the Star, is a Tangent to the Moon's Limb, or to the Horizon.

9. To make the Instrument useful, the Telescope ought to take a large Angle: And to make the Observation true, let the Star touch the Moon's Limb, not on the Outside of the Limb, but on the Inside.

As Halley and/or Newton would have immediately discovered, the 'minute and 1/6 minute' (360 per degree) divisions as described in paragraph 1 occupied far too much space on the base plate and added 40% to the weight of a relatively heavy instrument designed to be hand-held. Halley's experience with using diagonal scales on observatory instruments would also have alerted him to the practical difficulties in determining the point of shallow intersection between index and the long scribed diagonal line to the nearest 360th division even using a magnifier. Far easier to mark only 60 single divisions each representing a whole minute of arc on the index, at the same time shortening the length of each diagonal line (as illustrated in figure 13, chapter 30); thus widening the intersecting angles to highlight the intersection point as well as allowing for more weight to be cut from the base plate.

B. The 5 serious errors.

1. No diagonal scale (mentioned in paragraph 1 of Newton's note) is shown on either the artist's drawing or the engraving (figure 12, chapter 30). This omission implies that angular measurements would be limited to about ¼° at best.
2. Because both specula are set at 45° (paragraph 3 of Newton's note and as shown in figure 12 and both expanded sections below), one masks the other when attempting to take angular readings over about 50°. Newton's experimental design was clearly intended to measure small angles (as was Godfrey's design; figure 11, chapter 29) and yet the engraving (figure 12) allows for angles up to 90° implying Newton was unfamiliar with the basic principles of reflected light.

Three other errors are highlighted in the 2 sectional diagrams (below).

3. Index arm pivot point C is not in line with the measuring edge of the index arm on the engraving (but is on the artist's drawing). Because of this, accurate readouts on the rim scale would be impossible.
4. Specula G and H are not parallel when set at zero (but they are on the artist's drawing). This creates flaws in the angular readout and prevents self-checking.
5. Newton did not bother to state the obvious but the reflective surface of Specula H had to be centred directly over the pivot point C as in both Hadley designs (figures 9 & 10, chapter 29). The artist and engraver both position this mirror elsewhere and again imply Newton was unfamiliar with the basic principles of reflected light.

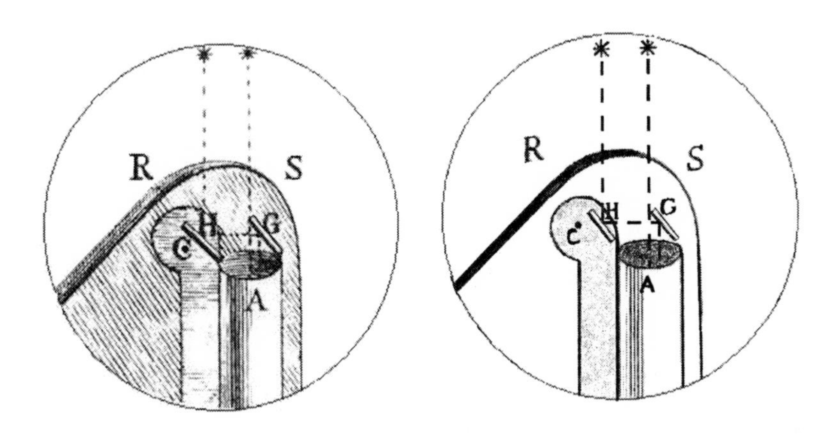

Section of engraving (figure 12) [1]. *Section of artist's drawing* [2].

Jointly these 5 mistakes serve to portray Newton's double-reflection principle as the product of a layman lacking any real knowledge of the subject.

However there are other points which Newton did not go into in detail which the artist and engraver took it upon themselves to presume. The scale runs from 0° to 45° on artist's version making no allowance for the mirrors doubling the angle measured but is shown correctly (from 0° to 90° on a 45° arc) on the published engraving (figure 12). Nevertheless zero is placed at wrong end of scale so the instrument has to be used upside down and thus the mirrors mask the images at even smaller angles. The fixed speculum G is positioned to intercept the bottom half (top half if the instrument is held upside down) of the light. This should intercept the outside half as in Halley's instrument (figure 5, chapter 13) and Hadley's 2 instruments (figures 9 &10). Speculum (mirror) G is not in line with the centre of the pivot C (off-centre to one side in drawing and to the other side in engraving) although the position is clearly mentioned in paragraph 4 of Newton's note.

(1) Isaac Newton, *A true Copy of a Paper found, in the Hand Writing of Sir Isaac Newton..* (R.S. Phil. Trans., 1742), Vol. 42, pp. 155-6.
(2) After Royal Society ink and wash drawing. RS Letters and papers 1742, Vol. 1, p 120, 1742.

APPENDIX 11

JOHN HADLEY'S MARINE QUADRANT PATENT.

The first section of John Hadley's patent document, which granted him patent protection in England, Wales and the town of Berwick-upon-Tweed for 14 years. The original was handwritten but, like other early patents, was later set in type and printed in 1857.

A.D. 1734 .... N°· 550.

Quadrant.

HADLEY'S PATENT

GEORGE THE SECOND, by the grace of God, &c., to all to whom these presents shall come, greeting.

WHEREAS our trusty and wellbeloved JOHN HADLEY, of East Barnet, in the county of Hertford, Esquire, hath by his petition humbly represented unto us, that he has, by long study and application, and at a very great expense, invented and contrived "AN INSTRUMENT OR QUADRANT FOR TAKING AT SEA THE ALTITUDE OF THE SUN, MOON, OR STARS AS ALSO ANY OTHER ANGLES, BY MEANS OF TWO SUCCESSIVE REFLECTIONS OF ONE OF THE OBJECTS FROM TWO SEVERAL PLATES OF LOOKING GLASS OR POLISHED METAL, BY WHICH THAT OBJECT IS BROUGHT TO COINCIDE WITH THE OTHER; AND THAT AN EXPERIMENT HAS BEEN MADE OF THIS INSTRUMENT IN A SMALL VESSEL, UNDER THE DISADVANTAGE OF A ROUGH SEA, WHEREIN THE OBSERVATIONS MADE THEREBY WERE FOUND MUCH MORE CONSISTENT THAN WHAT ARE USUALLY SEEN IN THOSE MADE ON BOARD A LARGE SHIP IN SMOOTH WATER, AND BY CONSEQUENCE MAY BE OF SIGNAL USE TO OUR ROYAL NAVY, AND TO THE TRADE AND NAVIGATION OF OUR KINGDOMS.........

Note. The 'experiment' refers to the seriously flawed *Chatham* tests.

NOTES

On Dates, Personages, Money and Reading Matter.

Dates.

The period covered by this book used two different dating systems; the old style (o.s.) Julian calendar in England whose year began on 25th March and the modern new style (n.s) Gregorian (beginning each year on 1st January) throughout much of Europe. The Gregorian calendar was 10 days in advance in the 17th century and 11 days from 19th February 1700. In 1752 when the Gregorian system was eventually introduced into England, Wales and Ireland (Scotland had sensibly but confusingly been using the new system since 1600) and her overseas possessions then including the American colonies, some people thought they had been robbed of 11 days of their lives. Even more confusing was the fact that astronomers and many scientists in England began each year on 1st January whilst retaining the old Julian dates; a scientific article published in February would carry the Julian date but the Julian and Gregorian year - 1705/6 for example.

The dates of birth and death of one of the main characters in this story, Isaac Newton, highlight this confusion. He was born on Christmas Day 1642 (o.s.) but 4th January 1643 (n.s.). He died on 20th March 1726/7 (o.s.) or 31st March 1727 (n.s.). When Edmond Halley arrived in Madeira (chapter 17) on 16th December 1698 (o.s.), the entire population of the town of Funchal was still celebrating Christmas, Portugal having adopted the Gregorian calendar in 1582. When he eventually managed to buy casks of wine on the 20th the locals were preparing to see in the New Year. Incidentally the legislators appeared to have forgotten to include the Hebridean island of St. Kilda whose 100 or so inhabitants finally abandoned the old Julian calendar in 1912, only 19 years before being evicted.

Wherever possible dates have been converted to the modern in order to readily link with SkyMap © or similar computer software and to limit general confusion.

Personages.

Monarchs.

Charles II, 1630-1685. Eldest son of Charles I and grandson of James I. Reigned 1660-1685.

James II, Duke of York 1633-1701.Younger brother of Charles II. Reigned 1685-1688.

Mary II, 1662-1694. Daughter of James II. Reigned jointly with William III 1689-1694.

William III, 1650-1702. Grandson of Charles I. Reigned 1689-1702.

Anne, 1665-1714. Daughter of James II and younger sister of Mary II. Reigned 1702-1714.

George I, 1660-1727. Grandson of James I. Reigned 1714-1727.

George II, 1683-1760. Son of George I. Reigned 1727-1760.

George III, 1738-1820. Grandson of George II. Reigned 1760-1820.

Main characters in date of birth order.

Wren, Sir Christopher., Fellow of the Royal Society. 1632-1723.

Hooke, Robert., FRS. 1635-1703.

Newton, Sir Isaac., FRS. 1643-1727.

Flamsteed, John., FRS. 1646-1719.
Shovell, Sir Cloudesley. 1650-1707.
George, Prince of Denmark., FRS. 1653-1708.
Halley, Edmond., FRS. 1656-1742.

Money.

In 1665, £1 was worth well over 200 times its current value based on the wages of a labourer, but by 1740 inflation had reduced this somewhat. No precise comparisons are possible and the occasional bracketed figures are simple 200x multiplications.

Reading matter.

A few of the books and articles cross-referenced in the text and listed at the end of each chapter are now out of print or difficult to access. However whenever possible modern accessible publications have been cited in preference to the original references and these are either still in print or readily obtainable through a local library. Some Royal Society papers cited are accessible either through the RS or the Newton Project web sites. Parliamentary Acts mentioned in the Epilogue are held in the reference sections of many major UK public libraries.

Andrewes, William, Ed., *The Quest for Longitude*, Harvard U.P., 1996.
Baily, Francis, *An account of the Revd. John Flamsteed,* Dawsons, London, 1966.
Bennett, Jim; Cooper, Michael; Hunter, Michael; Jardine, Lisa, *London's Leonardo; The Life and Work of Robert Hooke*, Oxford U.P., 2003.
Chapman, Allan, *Dividing the Circle; the development of critical angular measurements in astronomy, 1500-1850,* Horwood, Chichester, 1990.
Cook, Alan, *Edmond Halley; Charting the Heavens and the Seas,* Clarendon Press, Oxford,1998.
Craig, John, *Newton at the Mint,* Cambridge U.P., 1946.
Davies, K.G., *The Royal African Company,* Longmans, London, 1957.
Harris, Simon, *Sir Cloudesley Shovell; Stuart Admiral,* Spellmount, Staplehurst, 2001.
Inwood, Stephen, *The Man Who Knew Too Much. The Strange and Inventive life of Robert Hooke 1635-1703,* Macmillan, London, 2002.
Jardine, Lisa, *On a Grander Scale; The outstanding career of Sir Christopher Wren,* Harper Collins, London,. 2002.
MacPike, Eugene, *Correspondence & Papers of Edmond Halley,* Clarenden Press, Oxford, 1932.
Ronan, Colin A., *Edmond Halley; Genius in eclipse,* Macdonald, London, 1970.
Thrower, Norman J.W., Ed., *The Three Voyages of Edmond Halley in the Paramore 1698-1701,* The Hakluyt Society, London, 1981/1999.
Westfall, Richard, *The Life of Isaac Newton,* Cambridge U.P., 1999.
White, Michael, *Isaac Newton; The Last Sorcerer,* Fourth Estate, London, 1997.
Williams, J.E.D., *From Sails to Satellites; The Origin and Development of Navigational Science,* Oxford U.P., 1994.
Worsley, F.A., *Shackleton's Boat Journey,* Pimlico, London, 1999.

GLOSSARY

Some basic navigational explanations.

Latitude.

The imaginary horizontal bands that gird the Earth are termed 'lines of latitude'. Zero degree latitude describes the longest hoop that runs round the fattest middle part of the Earth; the equator. All places on this band are 'at' the same latitude - 0°. Other imaginary horizontal bands run all the way up to either pole from the equator like the hoops of a barrel, diminishing in girth until those at the poles (90°N- north, or S-south; or a quarter of a circle away from the equator) are mere pinpricks in the snow. Greenwich, across the River Thames from the 2012 Olympic Games site in East London is on the imaginary band designated 51°28'N-north of the equator (and thus 38°32' south of the North Pole). Dortmund in Germany or Kicking Horse Pass in Canada or any other place on that same horizontal band can be described as being 'on the same latitude' as Greenwich.

The Sun and stars have been used to determine approximate latitude for over 2000 years.

By referring to a solar almanac latitude can be established on any day of the year by measuring the height of the noonday Sun providing it is visible and its angle above the horizon can be measured reasonably accurately. But as with the magnetic compass, 17th century mariners knew the tables they relied on were suspect and were wise enough not to place too much reliance on their results. Many preferred to determine latitude by judging the height of the pole star which did not vary seasonally and did not require knowledge of the date. If *Polaris* was twinkling 31° above the horizon at any time during the night on any night of the year, your ship was within couple of degrees or so of the latitude of 31°N. But again there were snags, the most important being that once the equator was crossed the polar stars had disappeared out of sight beneath the northern horizon.

Longitude.

The word longitude describes the imaginary hoops or meridians that run vertically up and down the Earth joining the two poles. The 360 imaginary 1° hoops are, unlike the horizontal ones, all the same length and are widest apart at the equator like the segments of an orange. Convention now accepts that the longitudinal line running down through Greenwich is the one where numbering of these imaginary lines starts from; the prime meridian. Rather than run from 0° to 360° all the way round and back to Greenwich, these now run west and east to 180° where they meet round the other side. Luz in the Pyrenees and Tema in Ghana are both 'on' this same Greenwich prime 0° meridian.

Combining latitude and longitude in order to obtain a 'fix'; a cross on a chart which tells you exactly where you are.

Because the artificial horizontal lines of latitude and the vertical lines of longitude form a global grid, they can provide a unique locator reference which any navigator can recognise; the Kicking Horse Pass reference is 51°28'N., 116°23'W. for example. But although an explorer could determine the *latitude* of a bleak Kicking Horse Pass quite accurately when this story starts by measuring the Sun's noonday height or that of *Polaris* if he could somehow work out where the horizon was, he could have been in a leafy Greenwich park at the time where the readout would have been identical. Thus working out how also to determine *longitude* was the key to global navigation. Only then could a navigator obtain the combined locator set of figures and confirm that he was indeed in Canada rather than England. Whoever could work out a method of discovering how far round/along a latitudinal band he had travelled since departing from a known position such as Greenwich or Paris, would solve the entire global navigation problem, a problem that was, by the middle of the seventeenth century in *theory* entirely solvable by using any one of those three 'clocks' mentioned in the Introduction to carry Greenwich (or home port) time with you on your journey.

The lunar clock.

Because almost identical Moon-star positions are visible *at the same time* however far to the west or east of a known position the observer is, the Moon's 'slippage' of about ½° (its approximate diameter) per hour against the 'fixed' stars can be used like the hands of a clock. If in London the Moon is just touching a star at midnight London time, a very similar phenomenon will be seen 30° or 45° to the west at the same London time. You have successfully carried your (London time) clock with you. Then register the time of the Sun when it is at its highest at noon at your position out in the Atlantic to obtain 'local' time using an ordinary watch which can hopefully keep reasonable time for half a day, convert the difference between local and your London time into degrees at the rate of 15 per hour and you have both your longitude and your latitude (the Sun height at noon) in one fell swoop - in theory.

But the Moon is very close to a very large Earth observing platform and stars mind-bogglingly distant, so Moon-star alignments will change slightly when viewed from a different position on the Earth's surface and this (parallax) must be taken into account. Allowances also have to be made for the Moon's 223 cycle of elliptical orbits which means its apparent diameter, its speed and its distance from Earth are all constantly changing throughout each 18+ year period.

Lunar occultations and appulses mentioned in the text apropos the lunar clock.

The term occultation refers to a visually uncommon event when the Moon passes directly in front of a star, obscuring the latter from view for anything up to an hour depending on which part of the Moon blocks the starlight. As the Moon never strays far from the ecliptic (the apparent path followed by the Sun as it moves across the sky), the vast majority of stars in the heavens can never pass behind the Moon. Because of the Moon's continually changing orbit, a star might be obscured only once in a dozen years and then exceptionally in successive ones. An appulse event occurs at the point in time when the Moon comes close to a star without passing in front of it.

Magnetic variation.

A magnetised needle points to magnetic north which is nowhere near the geographic (true) North Pole. So the angular difference between true north and magnetic north varies considerably depending on one's location. To complicate matters further the magnetic poles are constantly, but slowly on the move. So in order to use a compass to steer by, this difference (variation) between true north and magnetic north has to be known and allowed for. No problem in the 21st century; your exact position is known (via the global positioning system - GPS or Sat-Nav) and the local magnetic variation information is printed on your chart. But in the 17th century magnetic variation could only be ascertained by checking the compass against known astronomical positions; the least reliable method being to point a magnetic compass in the direction of the pole star and estimate how many degrees to east or west the needle pointer was off line.

Angle measuring instruments mentioned in the text.
An astronomical octant is capable of measuring angles to a limit of 1/8th of a circle (45°), a sextant 1/6th (60°) and a quadrant 1/4th (90°) and all have been in use in observatories for over 2000 years. However with the introduction of marine angle-measuring devices which incorporated two mirrors, definitions became confusing because these effectively doubled the angular range. A twin-mirrored marine or surveying octant/sextant whilst still only possessing an arc of 45°/60° can measure angles as wide as 90°/120°.

Chronometer rating.
The 'rate' mentioned in various chapters refers to the number of seconds that a timekeeper is found to gain or lose every 24 hours when tested against a reliable source. Provided that this rate remains constant, allowance can be made and the chronometer remains 100% effective. Only if the rate fluctuates will discrepancies arise.

ACKNOWLEDGEMENTS

For supplying material, advice and assistance, I am particularly indebted to George Huxtable, Arthur Lennox, David Low, Graeme Robertson, John Young and my wife Anne. I would also like to thank Christine Woollett and the Royal Society Library staff and the staffs of the British Library, Cambridge University Library, the National Maritime Museum and the Public Records Office.

INDEX

References refer to chapters not pages.
Names of Ships & publications in *italics*.
E = Epilogue.
G = Glossary.
N = Notes.
A = Appendix & number.
Fig. = figure & number.

T

U

CPSIA information can be obtained at www.ICGtesting.com
Printed in the USA
LVOW041613271212

313476LV00004B/593/P